# Control of Underactuated Mechanical Systems

## Stabilization and Limit Cycle Generation

**Afef Hfaiedh**
University of Tunis EL Manar
National Engineering School of Tunis
Laboratory of Robotics, Informatics and Complex Systems
Tunis, Tunisia

**Ahmed Chemori**
LIRMM
University of Montpellier
CNRS
Montpellier, France

Series Editor
**Quan Min Zhu**



ELSEVIER

ACADEMIC PRESS
An imprint of Elsevier

Academic Press is an imprint of Elsevier
125 London Wall, London EC2Y 5AS, United Kingdom
525 B Street, Suite 1650, San Diego, CA 92101, United States
50 Hampshire Street, 5th Floor, Cambridge, MA 02139, United States



ISBN: 978-0-443-24020-1

For information on all Academic Press publications
visit our website at https://www.elsevier.com/books-and-journals

*Publisher:* Matthew Deans
*Acquisitions Editor:* Sophie Harrison
*Editorial Project Manager:* Thangminlun Kipgen
*Production Project Manager:* Prem Kumar Kaliamoorthi
*Cover Designer:* Mark Rogers

Typeset by VTeX

*To my father and mother for their unwavering support, love, and comfort in both good and bad times, and for bringing joy to my soul. I am eternally indebted to them.*

*To my brothers, who always believed in me and cared for me; this work represents the fruit of the sacrifices you made for my education.*

*To my niece, who brightens my days.*

*Afef Hfaiedh*

*To the memory of my parents*

*To my wife Chafia*

*To my kids Youssef, Sarah, and Yassine*

*Ahmed Chemori*

# Contents

# List of figures

# About the series editor

**Quan Min Zhu** is a Professor of Control Systems at the School of Engineering, University of the West of England, Bristol, UK. He earned his MSc from Harbin Institute of Technology, China, in 1983, and his PhD from the Faculty of Engineering, University of Warwick, UK, in 1989. With a primary research focus on nonlinear system modeling, identification, and control, Professor Zhu has authored over 300 papers in this domain, edited numerous influential books for Springer and Elsevier, and provided consultancy to a variety of industries.

# About the authors

## Afef Hfaiedh

Afef Hfaiedh received her Master's degree in Automatic Control, Robotics, and Signal Processing in 2015 and her Ph.D. degree in Automatic Control in 2021, both from the National Engineering School of Carthage. In 2023, she joined the Laboratory of Robotics, Informatics, and Complex Systems (RISC-LAB) as a postdoctoral researcher. She is currently working as a part-time professor at the University of Tunis El Manar, National Engineering School of Tunis, teaching engineering courses in electrical engineering. Her research interests include nonlinear, adaptive, and sliding-mode control and their applications in robotics and nonlinear underactuated mechanical systems.

## Ahmed Chemori

Ahmed Chemori received M.Sc. and Ph.D. degrees both in Automatic Control from the Grenoble Institute of Technology, Grenoble, France, in 2001 and 2005, respectively. During the year 2004/2005 he was a Research and Teaching assistant at Laboratoire de Signaux et Systèmes (LSS – Centrale Supelec) and the University Paris 11. Then, in 2006, he joined Gipsa-Lab (Former LAG) as a CNRS postdoctoral fellow. He is currently a senior Research Scientist in Automatic Control and Robotics with LIRMM, University of Montpellier, CNRS. His research interests include nonlinear (robust, adaptive, and predictive) control and their real-time applications in complex robotic systems.

# Preface

The history of control systems dates back to ancient times, with early examples such as the first water clock, or clepsydra, attributed to Ctesibius of Alexandria around 270 BC that used a float to regulate the water level in a reservoir. A similar concept was later illustrated in the peacock-shaped fountain designed by Al-Jazari. Now, automated systems are controlled by computer programs that execute algorithms known as control laws. The design of control laws that enable the execution of complex tasks or stabilize naturally unstable systems is a rapidly growing area of scientific research. It brings together several communities of researchers: mathematicians, physicists, mechanical engineers, roboticists, and more.

Improving and designing new control methods for mechanical systems has made it possible to use increasingly complex systems. Robotic systems, such as humanoid robots designed for service, are on their way to becoming part of daily human life. Controlling humanoid robots presents a real challenge in the field of mechanical systems control. First, these systems exhibit highly nonlinear dynamics. In other words, the differential equations that describe the system's movements over time, as well as the forces and torques generated by the system's actuators, contain nonlinear terms. Moreover, these systems are typically underactuated, meaning they have fewer actuators than degrees of freedom. As a result, they are usually nonlinear, characterized by non-integrable dynamic constraints. Such systems are, therefore, generally non-holonomic. In addition to their nonlinearities, humanoid robots must often interact with unpredictable environments and perform tasks that require a high degree of flexibility and precision. This further complicates the control design and demands sophisticated approaches to ensure stability and robustness in dynamic and uncertain environments. The combination of these challenges makes the control of humanoid robots, which serve as a case study of underactuated systems, a significant interdisciplinary research area in control theory.

Underactuation can occur in various situations, such as in vehicles like airplanes, helicopters, underwater robots, and similar systems where the number of control inputs is fewer than the degrees of freedom. This limitation introduces challenges in controlling all the system's movements directly. Underactuation can also be intentionally introduced during the design process to reduce the system's weight, complexity, and production costs, which can be beneficial in applications where simplicity and efficiency are key. Furthermore, underactuation can arise unexpectedly due to actuator failures or malfunctions. In such cases, a previously fully actuated mechanical system becomes underactuated, inheriting the same constraints and control difficulties. This makes the system more difficult to stabilize and control, requiring advanced control strategies to compensate for the loss of actuation and maintain performance.

This book deals with the challenging topic of control of underactuated mechanical systems. It addresses two main problems: the first is stabilization around an equilibrium point, and the second is the generation of stable limit cycles. The two concepts are important and can be illustrated with real-world examples. For instance, stabilization contributes to the aircraft's control by maintaining balance and preventing unwanted deviations from the flight trajectory. Smilarly, stable limit cycle generation is also significant in robotics, such as biped walking or the jump cycle of a Kangaroo hopping robot.

In the first part of this book, the first chapter introduces the reader to the concepts of stabilization and limit cycle generation. It serves to provide an intuitive understanding of the underlying principles, making it easier for the reader to grasp their significance in various applications of underactuated systems. The focus then progresses to the proposed system as a case study: namely, the inertia wheel inverted pendulum. The second chapter includes includes a thorough description of the system's dynamics, mathematical modeling, the use of the inertia wheel in many stabilization applications, and a detailed experimental setup, allowing readers to appreciate both theoretical and practical aspects.

The second part is devoted to the problem of stabilization, where a range of control solutions are presented and discussed. Each solution is analyzed in terms of its strengths and weaknesses. This section also includes comprehensive validation through numerical simulations and real-time experiments, demonstrating how the proposed strategies effectively achieve stabilization of the inertia wheel inverted pendulum.

The last part deals with some control solutions dedicated to the problem of stable limit cycle generation. The first solution is based on partial feedback linearization and dynamic control for optimal periodic reference trajectory tracking. The second control solution is a nonlinear model predictive control technique for a hopping robot. The last solution is a dual model-free control technique to deal with non-minimum-phase systems for stable limit cycles generation. All the proposed solutions are validated through numerical simulations or real-time experiments.

Part I

# General context and case study

Chapter 1

# Introduction and general context of underactuated mechanical systems

**Contents**



## 1.1 Introduction

The last decades have seen an increasing number of studies in the field of mechanical systems (MSs). The notion of "Degrees of Freedom (DOFs)" of a MS is commonly employed to denote the minimum number of independent variables that fully define its configuration. Determining those variables is crucial for ap-

Control of Underactuated Mechanical Systems. https://doi.org/10.1016/B978-0-44-324020-1.00008-7

propriate modeling of complex systems [1]. In addition, the degree of freedom concept gives rise to a basic classification of MSs. In fact, based on the number of DOFs and the number of control inputs, it is possible to classify MSs into three categories.

In this context, a MS that has fewer independent control inputs than DOFs is known as an underactuated mechanical system (UMS). This implies that, at least, one DOF is not actuated. The underactuation may be present due to the system dynamics, for instance, in the cases of underactuated aerial vehicles (UAVs) and vertical take-off and landing (VTOL) aircraft. In addition, various low-order nonlinear UMSs are also designed and presented in the literature for studying and analyzing purposes. There is no doubt that the complexity of their controller design is of considerable interest to researchers in the control community.

The underactuation can be intentionally introduced by the designer to minimize the cost and the weight and to reduce the energy consumption, which is ideal for satellites and aerospace systems. It may also exist accidentally in the case of an unexpected actuator failure in the runtime. Whatever the reasons that lead to the underactuation, this class of systems becomes the subject of active scientific research as a result of its diverse applications across different disciplines and of the challenges that arise, especially at the control level. There are interesting control issues in this class that require fundamental nonlinear approaches to deal with (i) the nonlinear dynamics, (ii) external disturbances, (iii) and variations in system parameters that may induce significant uncertainties. The existing mechanical systems are grouped into three different categories: fully actuated, redundant, and underactuated mechanical systems. A general description of each category is presented. This chapter will then provide an overview of some challenges of UMSs followed by some real applications of UMSs. The last section presents a literature review about some developed control approaches. Each control strategy is discussed briefly, shedding light on both the positive and negative aspects.

## 1.2 Classification of mechanical systems

For a century, various mechanical systems have been designed and have resulted in fruitful studies. Today, mechanical systems are part of many aspects of modern life. This variety of systems gained considerable interest, particularly within the control and robotics community. They may be classified in several ways, including, but not limited to:

- by field (industrial, aeronautical, surgical, spatial) systems;
- by structure (serial, parallel robot, underwater, flexible, humanoid, exoskeleton);
- by number of DOFs (3-DOF, 6-DOF);
- by number of inputs and outputs (SISO, MIMO,...).

In this section, an overview of the classification of MSs based on their degree of actuation is presented. We point to three categories, namely, fully actuated, underactuated, and redundant MS.

### 1.2.1 Fully actuated

Fully actuated mechanical systems (FAMSs) have the same number of independent actuators as degrees of freedom. Such a system is generally equipped with actuators positioned at the joints that enable it to follow the corresponding desired reference trajectory. Consider a general matrix form of a dynamic model that is expressed in the joint space by:

$$M(q)\ddot{q} + C(q,\dot{q})\dot{q} + G(q) = \tau$$

where

$q, \dot{q}, \ddot{q}$ denote the position, velocity, and acceleration vectors of the actuated joints, respectively;
$M(q)$ is the inertia matrix;
$C(q,\dot{q})$ represents the centrifugal and Coriolis forces matrix;
$G(q)$ denotes the gravitational forces vector;
$\tau$ is the input torques vector.

The right side of the dynamic model represents the total torque provided by its actuators. The relation between the generalized forces and the forces provided by the actuators is described by:

$$\tau = R(q)u \tag{1.1}$$

Consider that $m$ is the number of actuators and $n$ is the number of degrees of freedom, $u \in \mathbf{R}^m$ is the supplied torque, $R(q)$ denotes a matrix of dimensions $n \times m$ of the distribution of forces provided by the actuators on the generalized coordinates.

For fully actuated mechanical systems, the actuator force distribution matrix $R$ is square and invertible with $n = m$. Such a mechanical system is generally easier to control. Each joint is independently actuated and follows its desired reference trajectory. In general, a FAMS is feedback linearizable. In such a case, the control problem may be viewed as an equivalent problem of a linear system.

### 1.2.2 Redundant

Another type of mechanical system is called over-actuated or redundant systems. They are equipped with more actuators than configuration variables (DOFs). They have some merits that make them popular, especially in industry. These systems are more accurate and are generally faster and more rigid. Compared to fully actuated mechanical systems, the energy consumption of each actuator is reduced. Taking the example of redundant parallel robots, the working space is enlarged compared to their fully actuated counterpart and singularities may be limited with redundant actuation and thus improves their performances and kinematic capabilities. In the case of an emergency situation, over-actuation also allows an aircraft to successfully accomplish its tasks in case of an actuator breakdown, which improves their fault tolerance. However, control methods for this class are much more challenging to implement. Indeed, the complexity of the dynamics is increased as redundancy leads to more kinematic constraints. Moreover, antagonistic internal forces (also called internal prestress) may be induced and may harm the mechanical structure of the system. Those forces that exist in redundant parallel robots are due, for instance, (i) to measurement errors [2] or, (ii) to the non-synchronization of actuators. For all of the above reasons, the actuation redundancy must be taken into account in the control design stage. Some examples of redundant mechanical systems studied in the literature include parallel robots [3,4], over-actuated flying systems [5], and a five rotary-link (5R) robot [6].

### 1.2.3 Underactuated

UMSs have fewer actuators than degrees of freedom. This category is rich from both the application and control perspectives. It includes, but is not limited to, mobile robots, spacecrafts, underwater vehicles, surface vessels, gyrostabilizer systems, helicopters, non-prehensile robots, space robots, and planar underactuated manipulators. UMSs have a number of advantages: reduced (i) weight, (ii) mass, (iii) volume, (iv) energy consumption, and (v) cost.

- The dynamics of the system may be underactuated by nature, for example, spacecraft, helicopters, underwater vehicles, and wheeled mobile robots. Considering the example of a typical quadrotor, this possesses four rotors with propellers that allow it to navigate in a configuration space of 6-DOFs [7].
- From a system design perspective, it is generally recommended to reduce the number of actuators. This can be regarded as an advantage over fully actuated and redundant systems. Indeed, those FAMSs are generally heavier, very expensive, and require more maintenance. For these reasons, underactuation may be introduced by design for reduction of the cost or for some practical purposes, for example, satellites with two thrusters.
- In the case of an actuator failure, a FAMS may be transformed into an UMS. In such a case, the original control strategy is not convenient to maintain the continuity of any task. It is important that a control algorithm be designed in advance, to guarantee the normal operation, the safety, and the security of the system [8].
- UMSs may be imposed artificially to create complex low-order nonlinear systems for the purpose of gaining insight into controlling high-order UMSs. Some examples of 2-DOFs UMSs, including the inverted pendulum cart, the inertia wheel inverted pendulum [9], and the Furuta system, are presented in Fig. 1.1, Fig. 1.2, and Fig. 1.3, respectively.

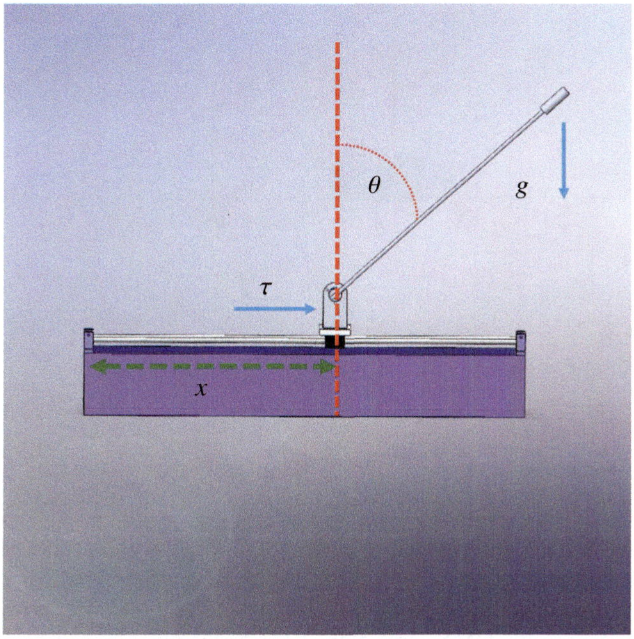


FIGURE 1.1 Inverted pendulum on a cart.

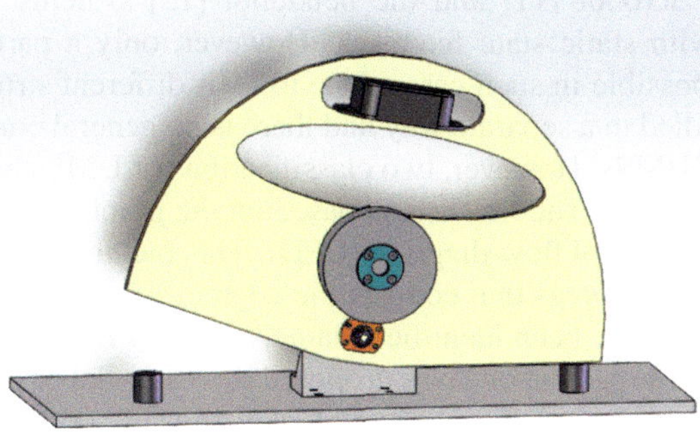

FIGURE 1.2 Inertia wheel inverted pendulum.

The restriction on the number of control inputs leads to several issues in the control design stage [10].

Many difficulties are often exhibited by such systems, such as their complex, highly nonlinear dynamics and their nonlinear coupling between actuated and

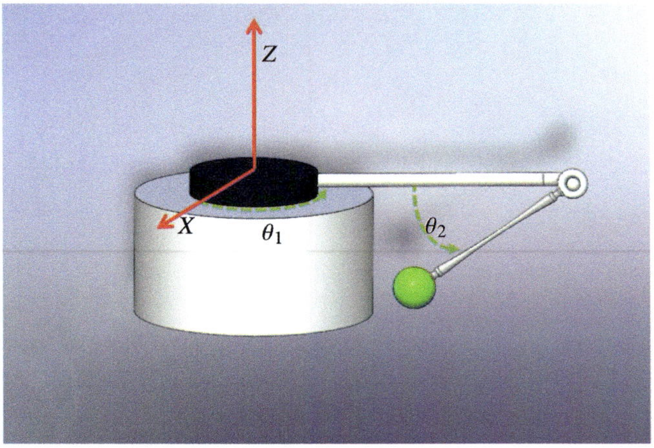


**FIGURE 1.3** Furuta system.

unactuated coordinates. By virtue of underactuation, non-integrable and second-order non-holonomic constraints can exist in the dynamic model of UMSs. For example, the dynamics of the pendubot is subject to second-order non-integrable differential constraints. UMSs are often characterized by a non-minimum-phase behavior due to the instability of their internal dynamics. As a result, a feedback controller can ensure a perfect tracking on part of the coordinates, but the instability of the system is still not reflected in the output. One of the control challenges is the lack of feedback linearizability. For instance, the dynamic model of the acrobot [11] and the pendubot [12] systems are not feedback linearizable with static state feedback. However, only a partial feedback linearization is possible in such cases. Due to their different structural properties, UMSs are studied in a separate way and there is no general control approach for all variants of UMSs. However, two classifications of UMSs have been reported in the literature [13]. The first one is based on the graphical characterization of UMSs called a control flow diagram (CFD). This method allows representation of the interactions forces that couples the degrees of freedom of an UMS [14]. Three structures have been identified, namely chain structure, tree structure, and isolated point. The second one was proposed by Olfati-Saber [15]. His main contribution consists of determining an explicit change of coordinates, based on the partial linearization, transforms an UMS into normal form. From the different obtained normal forms, we can categorize UMSs according to their structural properties and then design the appropriate control law for each class.

## 1.3 Why research underactuated mechanical systems?

In this chapter, we are mainly interested in control of UMSs. Conventional controllers applied for FAMSs cannot be straightforwardly implemented for UMSs. This is due to the fact that only a part of the dynamics can be controlled, the remaining part is known as the internal dynamics of the system. However, it is possible, by using nonlinear or hybrid techniques, to indirectly control the coordinates of the internal dynamics. Moreover, underactuation may result in degraded performance when subjected to external disturbances. Such a system requires the design of nonlinear control techniques to deal with external disturbances and parametric uncertainties. Within this section, various control challenges of UMSs are discussed.

### 1.3.1 First- and second-order holonomic constraints

The non-holonomic constraints in UMSs have received increasing attention over the past few years and represent a significant control problem [16]. Nonholonomic control systems are a class of nonlinear control systems that requires different control approaches from other nonlinear control systems. There exist two types of constraints. The first is called kinematic constraints, which geometrically limit the direction of movement. The second one is called dynamic constraints due to the dynamic balance at passive joints, where no torque can be applied [17]. A classification and comparative study of first-order and second-order constraints that arise from the conservation law and underactuation were addressed in [18]. Underactuation in some cases leads to non-holonomic constraints, such as non-integrable kinematics in the mechanical systems [19].

**Definition 1.3.1.** (Holonomic and non-holonomic constraints) Consider a system where $q$ is the vector of generalized coordinates

$$\ddot{q} = f(q, \dot{q}, u) \tag{1.2}$$

where $f$ is a function representing the dynamics of the system, $u$ is the input vector. If the system has constraints that limit motion and if some conditions can be written in the form of equations coupling the generalized coordinates and having the form

$$h(q, t) = 0 \tag{1.3}$$

then the constraint is said to be holonomic since it can be integrable.

#### *1.3.1.1 Example 1*

Consider the following system:

$$\dot{x}_1 = u x_2 \tag{1.4}$$

$$\dot{x}_2 = -u x_1 \tag{1.5}$$

where $[x_1, x_2]^T$ is the system state vector, $u$ is the input control signal. There exists a kinematic constraint described as follows:

$$x_1 \dot{x}_1 + x_2 \dot{x}_2 = 0 \tag{1.6}$$

This constraint is integrable and yields:

$$\frac{1}{2}x_1^2 + \frac{1}{2}x_2^2 = c \tag{1.7}$$

where $c$ is a constant. This constraint described in (1.7) is called a holonomic constraint. If the system is subjected to kinematic, non-integrable, and independent constraints in the form (1.3), the system is said to be non-holonomic. Generally, underactuation leads to the presence of nonlinear and non-integrable dynamic constraints. These systems are therefore said to be non-holonomic.

#### 1.3.1.2 Example 2

Consider the model of a mobile robot described by the following equations:

$$\dot{x} = v\cos\theta \tag{1.8}$$

$$\dot{y} = v\sin\theta \tag{1.9}$$

The kinematic constraints of the model are described by:

$$\dot{x}_1 \sin\theta - \dot{y}\cos\theta = 0 \tag{1.10}$$

This constraint is called non-holonomic, since it cannot be integrable or written in the form of a derivative with respect to the time of any other variables.

The difficulty in non-holonomic systems is generally considered at the level of the kinematic model, while it is related to the dynamics for underactuated systems.

Wheeled mobile robots [20,21], surface vessels, planar underactuated manipulators [17] are some examples in which the constraints are non-integrable or partially integrable. Based on a result from [22], a MS with first-order non-holonomic constraints cannot be asymptotically stabilized by a smooth static state feedback. A reduction of an UMS with non-holonomic first-order constraints and symmetry is possible, as in the case of the snakeboard system [15].

### 1.3.2 Nonlinear dynamics and coupled inputs

The nonlinear dynamics of UMSs make the control task difficult to design. For such complex systems, conventional linear control strategies may not guarantee the stabilization or trajectory tracking of the system. While nonlinear FAMSs are well established in control theory, nonlinear UMSs are known to be more challenging and difficult to deal with. Generally, no control objective will be achieved, if existing nonlinear control methods of FAMSs are straightforwardly applied to UMSs. Taking for instance, the tracking failure on a specific actuator of a FAMS, this will propagate to the others actuators, resulting in (i) a considerable increase of the nonlinearities in the dynamics, (ii) a loss of tracking, and (iii) possible damage of the mechanical structure. In this case, it is important to consider the nonlinear dynamic coupling between actuated and non-actuated coordinates and to switch to another pre-implemented control algorithm, in order to guarantee the normal operation of the system.

The challenging part is that the controller design differs depending on the position of the damaged actuator in the whole structure, while considering the nonlinear dynamics of the system. This leads to a preliminary analysis of various different cases.

The complexity of the dynamic model usually increases significantly, depending on several factors. They may include, a high degree of underactuation, the presence of certain constraints, singularities, frictions, etc. Moreover, the mechanism of UMSs (serial, parallel, flexible, etc.) plays a significant role in this regard. It is well known that systems with a closed chain structure are more difficult in terms of analysis, modeling, and control design, compared to those with an open chain structure.

By considering the underactuation in such a case, the control design will be much more difficult. For the above reasons, control of UMSs is often described in the literature as a very challenging task compared to FAMSs. They require effective analysis and new control strategies.

### 1.3.3 Non-minimum-phase system

The difference between the number of degrees of freedom $n$ and the number of independent actuators $m$ of an underactuated mechanical system is called the degree of underactuation defined as $n-m$. Let us consider the vector of generalized coordinated, $q=[q_a^T, q_{na}^T]^T$, where $q_a^T \in \mathbf{R}^n$ and $q_{na}^T \in \mathbf{R}^{n-m}$ represent the actuated and non-actuated coordinates, respectively. We can rewrite the system $M(q)\ddot{q}+C(q,\dot{q})\dot{q}+G(q)=R(q)u$ in the following form:

$$\begin{aligned} M_{11}(q)\ddot{q}_a + M_{12}(q)\ddot{q}_{na} + C_{11}(q,\dot{q})\dot{q}_a + C_{12}(q,\dot{q})\dot{q}_{na} + g_1(q) &= \tau_a \\ M_{21}(q)\ddot{q}_a + M_{22}(q)\ddot{q}_{na} + C_{21}(q,\dot{q})\dot{q}_a + C_{22}(q,\dot{q})\dot{q}_{na} + g_2(q) &= 0 \end{aligned} \tag{1.11}$$

where

$M(q)=\begin{pmatrix} M_{11}(q) & M_{12}(q) \\ M_{21}(q) & M_{22}(q) \end{pmatrix}$ is the inertia matrix;

$C(q)=\begin{pmatrix} C_{11}(q,\dot{q}) & C_{12}(q,\dot{q}) \\ C_{21}(q,\dot{q}) & C_{22}(q,\dot{q}) \end{pmatrix}$ include the Coriolis and centrifugal components;

$G(q)=\begin{pmatrix} g_1(q) \\ g_2(q) \end{pmatrix}$ denotes the gravity vector;

$\tau_a$ of dimension $m$, represents the vector of the generalized force;

$u \in \mathbf{R}^n$, represents the vector of the control input.

The distribution of efforts matrix $R(q)$ on the generalized coordinates is of dimension $n \times m$. A mechanical system is said to be underactuated if the

$rank R(q)$ is less than $n$. We consider that the $rank R(q)$ is equal to $m$, meaning that the actuators of the system are independent and, moreover, that $m$ is less than $n$.

The system is then decomposed into two sub-systems, represented by the two equations represented in (1.11). The second equation can be seen as $n - m$ constraints associated with the first equation of (1.11) that any trajectory $q(t)$ must be satisfied at all $t$ in order to be tracked by the system. Let us linearize the first sub-system of (1.11) by state feedback. We start by solving the second equation of (1.11), to obtain:

$$\ddot{q}_{na} = -M_{22}^{-1}(q)(M_{21}(q)\ddot{q}_a + C_{21}(q,\dot{q})\dot{q}_a + C_{22}(q,\dot{q})\dot{q}_{na} + g_2(q)) \quad (1.12)$$

We substitute Eq. (1.12) in the first sub-system, to obtain:

$$M'(q)\ddot{q}_a + C_1'(q,\dot{q})\dot{q}_a + C_2'(q,\dot{q})\dot{q}_{na} + g'(q) = \tau_a \quad (1.13)$$

where

$$\begin{aligned} M'(q) &= M_{11}(q) - M_{12}(q)M_{22}^{-1}(q)M_{21}(q) \\ C_1'(q,\dot{q}) &= C_{11}(q,\dot{q}) - M_{12}(q)M_{22}^{-1}(q)C_{21}(q,\dot{q}) \\ C_2'(q,\dot{q}) &= C_{12}(q,\dot{q}) - M_{12}(q)M_{22}^{-1}(q)C_{22}(q,\dot{q}) \\ g'(q) &= g_1(q) - M_{12}(q)M_{22}^{-1}(q)g_2(q) \end{aligned}$$

We define the control law as:

$$\tau_a = M'(q)\nu + C_1'(q,\dot{q})\dot{q}_a + C_2'(q,\dot{q})\dot{q}_{na} + g'(q) \quad (1.14)$$

where $\nu$ is a new control input to be defined. The system (1.11) with (1.14) becomes:

$$\ddot{q}_a = \nu \quad (1.15)$$

$$M_{22}(q)\ddot{q}_{na} = -M_{21}(q)\nu - C_{21}(q,\dot{q})\dot{q}_a - C_{22}(q,\dot{q})\dot{q}_{na} - g_2(q) \quad (1.16)$$

The first sub-system in (1.15) is linearized. The second sub-system in (1.16) is called the internal dynamics of (1.11).

The zero dynamics is obtaining by substituting $\ddot{q}_a$, $\dot{q}_a$, $q_a$, and $\nu$ by zero in the equation $M_{22}(q)\ddot{q}_{na} = -M_{21}(q)\nu - C_{21}(q,\dot{q})\dot{q}_a - C_{22}(q,\dot{q})\dot{q}_{na} - g_2(q)$.

An underactuated mechanical system whose zero dynamics is not stable is called a non-minimum-phase system. Otherwise, if the zero dynamics is stable, it is considered to be a minimum-phase system. The latter can therefore be completely stabilized by designing a control law that stabilizes the actuated part. Unfortunately, an UMS is in the general case a non-minimum-phase system and therefore difficult to control.

### 1.3.4 Uncertainties and parametric variations

There are differences between a real system and the mathematical model, even with a good identification. They are referred to as uncertainties. Uncertainties can be caused by many factors, including, but not limited to, sensors and equipment, external contact force, modeling simplifications, unknown disturbance inputs, poor system identification, etc. In control theory, uncertainties are classified into two types. The first one is matched uncertainty that directly affects the control input. They can significantly impact system stability and performance. The second is called unmatched uncertainty, which does not directly influence the control input but can affect the system performance. As a result of these uncertainties, there may be an unexpected behavior of the system that affects its safety, increases tracking errors, and even instability of the closed-loop system. These consequences are well addressed for nonlinear FAMS, however, they may be harmful to highly sensitive UMSs. For the aforementioned reasons, robust control approaches such as sliding mode control (SMC) and adaptive control, active disturbance rejection, event-triggered control, among others are essential to handle such uncertainties in UMSs.

## 1.4 Stabilization problem

In robotics, some applications require stabilization techniques to achieve consistent and reliable performance, particularly in systems that are underactuated where maintaining balance and preventing instability are challenges. In this section, the concepts of stabilization are introduced and demonstrated through illustrative examples.

### 1.4.1 Concepts of stabilization

The control of the inverted pendulum generally consists of stabilizing the pendulum in an unstable equilibrium position, a vertical position of the pendulum around zero, and maintaining it in this position, even when it is affected by external disturbances or by parametric variations such as changes in the center of mass and payload. If we assume that the system is in some initial condition, as illustrated in Fig. 1.4. The objective is to find a control input $u$ to bring $x(t)$ to $x_e$ and maintain it around this point. This is called stabilization around an unstable equilibrium point. If we assume that the system is in some initial condition, as illustrated in Fig. 1.5, the main goal is to design a control law $u$ to bring and maintain it around an oscillating trajectory. This is a simple illustration of an orbital stabilization.

### 1.4.2 Basic ideas

Local stabilization at an equilibrium point consists of stabilizing the unactuated part of the system at its unstable equilibrium position, and the actuated part at

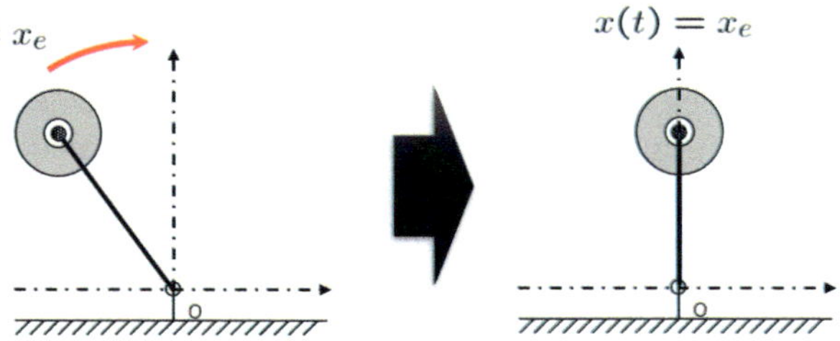


FIGURE 1.4 Problem of stabilization around an unstable equilibrium point.

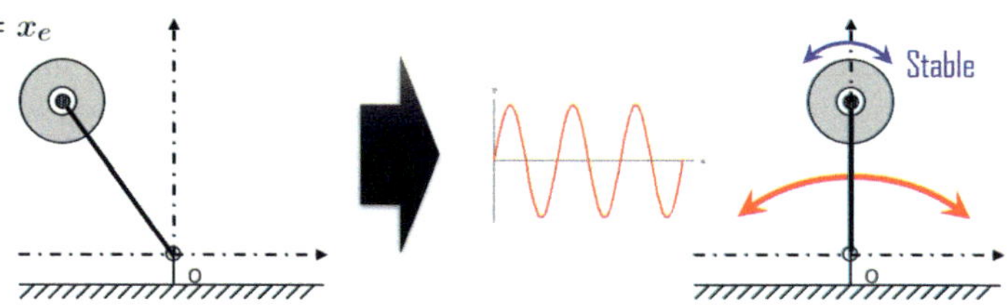


FIGURE 1.5 Problem of orbital stabilization.

a desired target position where the initial conditions are sufficiently close to the desired equilibrium position. Let us consider the system equation described in (1.17):

$$\dot{x} = f(x, u), u = \vartheta(x) \tag{1.17}$$

where $x$ denotes the state of the system and $\vartheta(x)$ is the control input. The local stabilization means that the origin of $\dot{x} = f(x, \vartheta(x))$ is asymptotically stable. When dealing with local stabilization, we usually proceed to the linearization of the system around the desired equilibrium followed by the synthesis of a linear static state feedback. However, the control, synthesized based on the linearized model, may have a restricted domain of attraction [106]. In addition to the presence of friction forces, it leads to low amplitude limit cycles, resulting in oscillations around the unstable desired equilibrium position.

Global stabilization involves bringing the unactuated part from any initial conditions to the unstable equilibrium position and the actuated part to a desired final position. In this case, the origin of $\dot{x} = f(x, \vartheta(x))$ is globally asymptotically stable.

Orbital stabilization consists of self-generation of a periodic orbit by the closed-loop system.

### 1.4.3 Illustrative example

The dynamic model of the Vertical Take-Off and Landing (VTOL) is described by:

$$\ddot{x} = \frac{1}{m}(-\sin(\theta)u_1 + \varepsilon\cos(\theta)u_2) \tag{1.18}$$

$$\ddot{y} = \frac{1}{m}(\cos(\theta)u_1 + \varepsilon\sin(\theta)u_2) - g \tag{1.19}$$

$$\ddot{\theta} = \frac{1}{I}u_2 \tag{1.20}$$

where $(x, y)$ represents the position of the vehicle's center of mass and $\theta$ its roll angle $m$, $I$, and $g$ denote the total mass of the vehicle, its inertia, and the acceleration due to gravity, respectively. $\epsilon$ denotes a positive coupling coefficient between the control input $u_2$ and the acceleration along the lateral axis of the vehicle:

$$\begin{pmatrix}\ddot{x}\\ \ddot{y}\end{pmatrix} = \frac{1}{m}\begin{pmatrix}-\sin(\theta) & \epsilon\cos(\theta)\\ \cos(\theta) & \epsilon\sin(\theta)\end{pmatrix}\begin{pmatrix}u_1\\ u_2\end{pmatrix} + \begin{pmatrix}0\\ -g\end{pmatrix} \tag{1.21}$$

If we choose the following linearizing state feedback control law:

$$\begin{pmatrix}u_1\\ u_2\end{pmatrix} = m\begin{pmatrix}-\sin(\theta) & \cos(\theta)\\ \frac{1}{\epsilon}\cos(\theta) & \frac{1}{\epsilon}\sin(\theta)\end{pmatrix}\left(\begin{pmatrix}v_1\\ v_2\end{pmatrix} + \begin{pmatrix}0\\ g\end{pmatrix}\right) \tag{1.22}$$

we obtain the partially linearized model of the system, which is described in the form:

$$\ddot{x} = v_1 \tag{1.23}$$

$$\ddot{y} = v_2 \tag{1.24}$$

$$\ddot{\theta} = K(\sin(\theta) + v_1\cos(\theta) + v_2\sin(\theta)) \tag{1.25}$$

where $v_1$ and $v_2$ are the control inputs. $K = \frac{m}{\epsilon I}$ is a constant. The third equation represents the internal dynamics. The fully linearized sub-system can be stabilized by an appropriate choice of control inputs $v_1$ and $v_2$. The stability analysis of the overall system will depend on the stability of the system's zero dynamics. By setting $v_1 = v_2 = 0$, we obtain:

$$\ddot{\theta} = K(\sin(\theta)) \tag{1.26}$$

Eq. (1.26) is simply the equation of a pendulum with length $K$. The phase plane for different initial conditions is shown in Fig. 1.6. The equilibrium states are thus $(\theta, \dot{\theta})_{p1} = (0, 0)$, which is unstable, and $(\theta, \dot{\theta})_{p2} = (\pi, 0)$, which is stable but not asymptotically stable.

Around the stable equilibrium point, there is a family of periodic orbits. Outside this family of orbits, along the $\theta$-axis, there are unbounded trajectories.

## 1.5 Stable limit cycle generation problem

In addition to the stabilization around unstable equilibrium points, the existence of limit cycles is an active research topic in the control of nonlinear UMSs. An

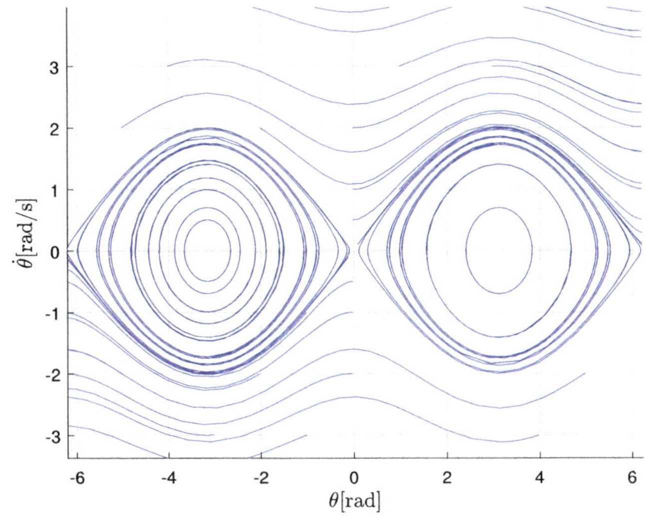


**FIGURE 1.6** Phase plane of the zero dynamics of the VTOL for $K = 1$.

attractive (stable) limit cycle represents a closed and isolated trajectory that, as time approaches infinity $t \to \infty$, is reached by all nearby trajectories. It may be semi-stable, which means that only some closed trajectories converge to it, as $t \to \infty$, or unstable, meaning that all nearby trajectories diverge from it, as $t \to \infty$ [39].

Limit cycles are in high demand in several applications, such as in the repetitive tasks of industrial manipulators, walking robots, non-prehensile robots. For example, the running motion or hopping motion is considered as a series of accurate repetitions of a closed trajectory considered as a limit cycle. The stability of a running underactuated biped robot is addressed in [40]. The system under consideration was equipped with only two legs and one actuator attached in the hip joint. The periodic running motion is composed of three consecutive phases, including the stance, the flight, and the transition phases. They describe the natural passive limit cycle of a running motion. Various studies have been reported in the literature concerning the generation of stable limit cycles for UMSs. Some methods are based on virtual holonomic constraints [41,42], bifurcation theory [43,44], parameterized polynomial functions [45,46], etc.

### 1.5.1 Definition

With the system's phase space, we can observe the evolution of the state of an underactuated mechanical system, where the trajectories are plotted for different initial conditions. The equilibrium configurations of the mechanical system are the fixed points on the phase space plot.

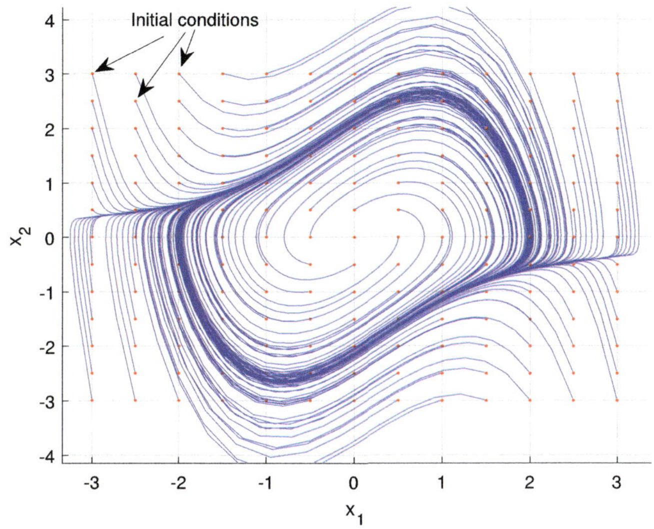


FIGURE 1.7 Phase plane of the Van der Pol equation $\varepsilon = 1$.

#### 1.5.1.1 Periodic trajectory

A cycle is represented in the phase plane by a closed curve called a periodic orbit or closed orbit. If we consider a pendulum subject to gravity and neglecting friction, the oscillations around the equilibrium point are cycles. We consider a cycle or periodic trajectory, a trajectory $q(t) \in \mathbb{R}^n$ if there exists a period $T \in \mathbb{R}$ such that

$$q(t+T) = q(t), \vee\ t \ \geq 0 \tag{1.27}$$

**Definition.** A limit cycle is an isolated cycle in the phase plane.

This explains that neighboring trajectories are not periodic, but approach or move away from the cycle. An illustration is the oscillator introduced by Balthazar Van der Pol, a Dutch electrical engineer. The equations are described as follows:

$$\dot{x}_1 = x_2 \tag{1.28}$$

$$\dot{x}_2 = \varepsilon(1 - x_1^2)x_2 - x_1 \tag{1.29}$$

Fig. 1.7 depicts the phase plane of the Van der Pol dynamics for different initial conditions.

For the pendulum, each orbit is periodic, however, the cycle of the Van der Pol oscillator is isolated because the neighboring orbits of this cycle are not periodic but approach the cycle. In this case, it is called a "limit cycle".

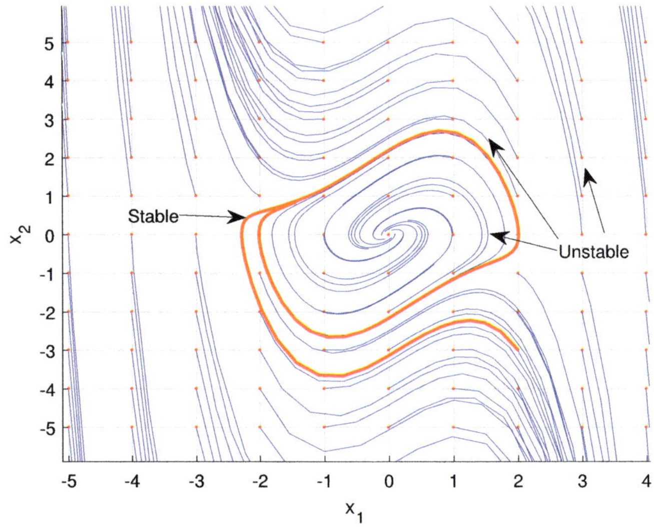


**FIGURE 1.8** Phase plan of reverse time Van Der Pol oscillator $\varepsilon = 1$.

## 1.5.2 Stability of a limit cycle

A limit cycle can be stable or unstable depending on the behavior of the neighboring orbits as time tends to infinity. A stable limit cycle is one where neighboring orbits approach the cycle as time tends to infinity. This means that if you start near the limit cycle, your trajectory will eventually converge to the limit cycle.

A limit cycle is called unstable when neighboring orbits do not approach the cycle as time $t$ tends to infinity. This means that if you start near the limit cycle, your trajectory will eventually diverge from the limit cycle. Let us consider the reverse time Van der Pol oscillator described by Eqs. (1.30):

$$\begin{aligned} \dot{x}_1 &= -x_2 \\ \dot{x}_2 &= -\varepsilon(1 - x_1^2)x_2 + x_1 \end{aligned} \tag{1.30}$$

Unlike the orbit in the previous Fig. 1.7, the neighboring orbits of this cycle do not approach any cycle (see Fig. 1.8). For different initial conditions, neighboring trajectories depicted in the phase plan diverge or tend toward zero.

## 1.5.3 Illustrative example of limit cycles

### 1.5.3.1 Pendulum

We consider a simple one-DOF actuated pendulum (see Fig. 1.9) subject to gravity, and air friction. A sinusoidal control input corresponding to a torque is applied on the rotation axis. The nonlinear mathematical dynamics of the

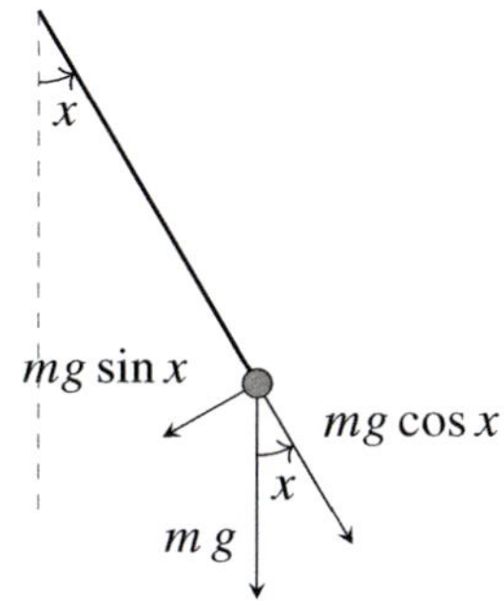


**FIGURE 1.9** The simple pendulum system.

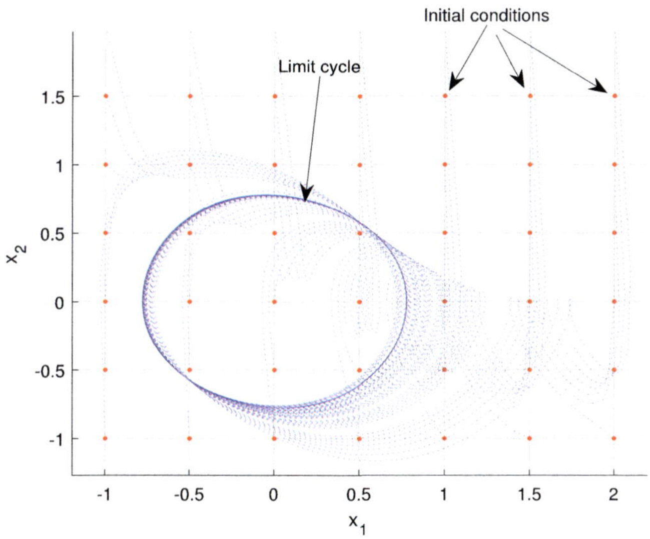


**FIGURE 1.10** Phase plane of the pendulum, $m=1$, $l=1$, $g=9.81$, $k=10$, $a=10$, $T=2\pi$, $x_1=x$, $x_2=\dot{x}$.

pendulum is then described as follows:

$$ml^2\ddot{x} = mgl\sin(x) - k\dot{x} + u \tag{1.31}$$

where $x$ denotes the angular position of the pendulum, $m$ represents the mass of the pendulum, $g$ is the acceleration due to gravity, $l$ denotes the length of the pendulum, $k$ is the coefficient of friction, and $u = a\sin(2\pi\frac{t}{T})$ is the torque applied to the pendulum, where $a$ and $T$ denote the amplitude and the period, respectively. Fig. 1.10 demonstrates the existence of a stable limit cycle as neighboring trajectories with different initials conditions reach the cycle.

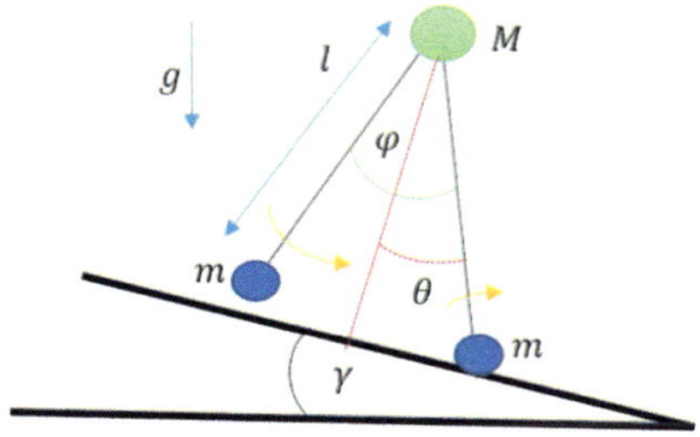


FIGURE 1.11 A passive walking illustration.

### 1.5.3.2 Limit cycle walking

For many years, researchers have been studying the replication of human walking with robots. Considering the bipedal walking in the case of a humanoid robot, even if it is fully actuated, it becomes underactuated. Indeed, if we consider the collision of the swing foot with the ground known as the impact phase, the system must then be described with six additional degrees of freedom corresponding to position and orientation in space. If we consider the case of the simplest model of passive walking [51] (see Fig. 1.11). The compass-gait biped system is illustrated by the two-link model. It is an idealized mechanical system that mimics the basic walking motion of a human. It consists of two rigid links representing the legs, connected at a hip joint. The first leg, named the stance leg, remains in contact with the ground, while the other, called the swing leg, swings forward during walking. The model exhibits passive walking, meaning it can walk down a slope using only gravity, without any active control or motors [52].

The motions of the stance leg and swing leg are generally periodic. We can observe in Fig. 1.12 that the walker will converge to a periodic motion where each step is nearly identical to the previous one. It represents the passive limit cycle trajectory of the compass gait.

The stability of these limit cycles can be influenced by initial conditions and the slope of the walking surface [52]. In this context, many research studies focus on the bifurcation of passive gaits, investigating how small changes in system parameters, such as slope angle or mass distribution, lead to different walking patterns or gaits [53–55].

## 1.6 Underactuation in a broad range of applications

In many practical systems, achieving a lightweight and energy-efficient design is important. For example, in aerial drones, reducing the number of control inputs can lead to simpler, more compact designs that consume less power. Underactuation is sometimes recommended in such cases. This approach can also help the system remain operational throughout the mission in the case of actuator fail-

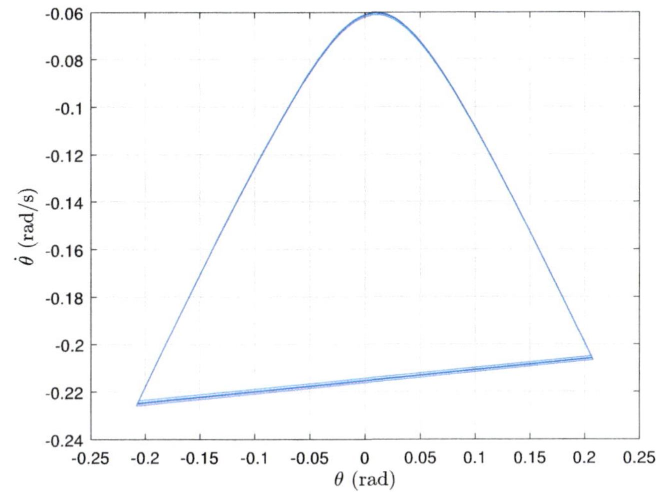


**FIGURE 1.12** Phase plot of $\dot{\theta}$ versus $\theta$.

ure, as seen in spacecraft or autonomous vehicles. In this section, several real applications of UMSs are presented.

### 1.6.1 Aerospace underactuated systems

Let us consider a particular underactuated mechanical system: the vertical take-off and landing (VTOL) aircraft depicted in Fig. 1.13 representing the system in the planar plane. The vehicle is modeled in the frontal plane $(xOy)$ during the take-off and landing phase, as introduced in [47]. The planar VTOL has three generalized coordinates, the position $(x, y)$ of the vehicle's center of mass and its roll angle $\theta$ in the $(xOy)$ plane. The control inputs are $u_1$ exerted by the engines in the center of mass directed upwards of the vehicle and the torque $u_2$ generated by the reaction control system (RCS) mechanism on the vehicle in the $(xOy)$ plane. The VTOL has three degrees of freedom for only two actuators, making it an UMS. According to the Lagrangian formalism, the kinetic energy $T$ and potential energy $V$ of the system are expressed as:

$$T = \frac{1}{2}m\left(\dot{x}^2 + \dot{y}^2\right) + \frac{1}{2}I\dot{\theta}^2 \tag{1.32}$$

$$V = ygm \tag{1.33}$$

where $m$ is the total mass of the vehicle, $I$ its inertia, and $g$ the gravitational acceleration. The equations of motion are derived by applying the Euler–Lagrange formula:

$$\frac{d}{dt}\left(\frac{\partial L}{\partial \dot{x}}\right) - \frac{\partial L}{\partial x} = m\ddot{x} = \tau_x \tag{1.34}$$

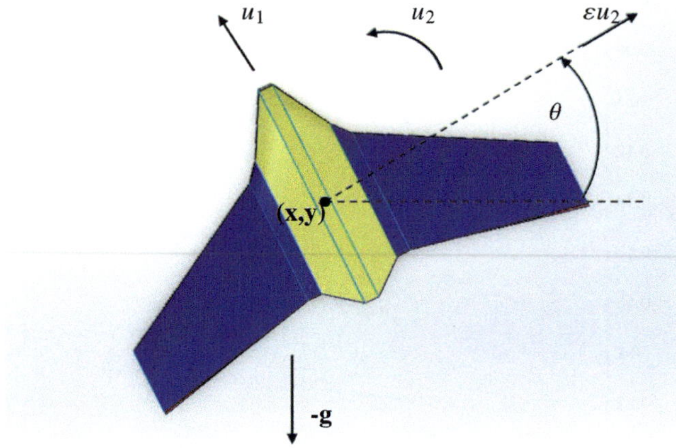


**FIGURE 1.13** Representation of the planar VTOL with its coordinates $x$, $y$, and its control inputs $u1$ and $u2$.

$$\frac{d}{dt}\left(\frac{\partial L}{\partial \dot{y}}\right) - \frac{\partial L}{\partial y} = m\,(\ddot{y} + g) = \tau_y \tag{1.35}$$

$$\frac{d}{dt}\left(\frac{\partial L}{\partial \dot{\theta}}\right) - \frac{\partial L}{\partial \theta} = I\ddot{\theta} = \tau_\theta \tag{1.36}$$

with $L = T - V$ being the Lagrangian of the system and $\tau_x$, $\tau_y$, $\tau_\theta$ the generalized forces. The control effort distribution matrix is as follows:

$$\begin{pmatrix} \tau_x \\ \tau_y \\ \tau_\theta \end{pmatrix} = \begin{pmatrix} -\sin(\theta) & \epsilon\cos(\theta) \\ \cos(\theta) & \epsilon\sin(\theta) \\ 0 & 1 \end{pmatrix} \begin{pmatrix} u_1 \\ u_2 \end{pmatrix} \tag{1.37}$$

where $\epsilon$ is a positive coupling coefficient between the control input $u_2$ and the acceleration along the lateral axis of the vehicle. Indeed, the torque exerted by the control system on the reaction forces induces a lateral force on the vehicle, directed in the opposite direction. This phenomenon is due to the slight downward inclination of the vehicle's wings: the reaction forces on the wings are not exactly opposite, resulting in a non-zero lateral resultant force.

The dynamic model of the VTOL is described by;

$$\ddot{x} = \frac{1}{m}\left(-\sin(\theta)u_1 + \varepsilon\cos(\theta)u_2\right) \tag{1.38}$$

$$\ddot{y} = \frac{1}{m}\left(\cos(\theta)u_1 + \varepsilon\sin(\theta)u_2\right) - g \tag{1.39}$$

$$\ddot{\theta} = \frac{1}{I}u_2 \tag{1.40}$$

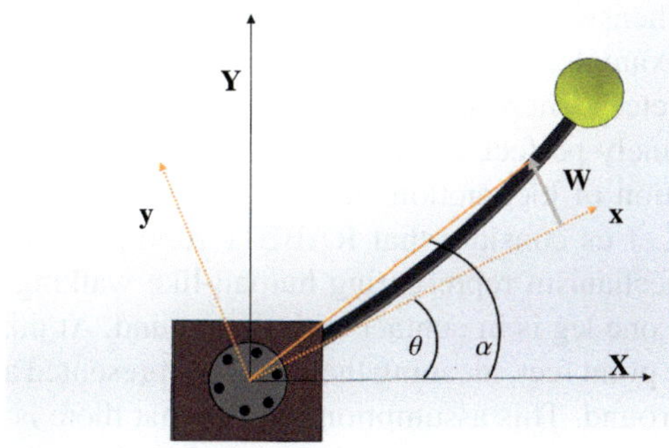


**FIGURE 1.14** Illustration of a 1-DOF flexible arm, $\theta$ is the joint variable, $W$ the deformation caused by the external force.

The complex nonlinear and coupled dynamics of a VTOL pose challenges in effectively implementing linear control techniques. To attain the desired performance, more advanced methods are required, as proposed in [48–50].

### 1.6.2 Flexible systems

Flexible-link robots are becoming increasingly popular in industrial and space robotics applications because they are lighter and more responsive than rigid robots [15]. The flexible arm can handle a larger payload relative to its own weight or size. It provides a rapid achievable movement, and reduced energy consumption. Due to its advantages, researchers have been studying the control of flexible robots for over ten years. The work in [56] addresses the trajectory control issue concerning a nonlinear one-link flexible arm, illustrated in Fig. 1.14. The authors suggest a state-feedback control strategy derived from the input–output inversion algorithm to precisely follow any specified smooth trajectory for the output. A study concerning the trajectory tracking and vibration suppression was proposed by [57] where a partial decoupling of the joint variable and elastic deformation is obtained by a nonlinear feedback control method, derived from the model. Then, the inverse dynamic method and Linear Quadric (LQ) control method is applied to deal with the trajectory tracking and vibration suppression. Another approach proposed by [58] is based on an iterative learning [59] that improves the tracking performance by exploiting repetitions of the desired task.

### 1.6.3 Locomotion systems

In recent years, scientists have been working to replicate specific functions of humans or animals through robotic prototypes. These studies have led to increasingly advanced robots equipped with artificial intelligence and numerous sensors to adapt to their environments. These are either humanoid robots, resembling humans, or bio-inspired robots. Indeed, mimicking a specific function requires a comprehensive study of the anatomy and physiology of the specific mechanism (for example, the structure of the legs for walking, the hand for grasping objects, etc.). These mechanisms define the human or animal structure, which is divinely perfect and extremely complex. Among these complex systems, the function of locomotion stands out, particularly for its underactuation properties. Let us consider that RABBIT robot, which was designed to be the simplest mechanism representing human-like walking [60]. During the swing phase, only one leg is in contact with the ground. At this stage, the robot is assumed to have point feet, meaning the feet are represented as single points of contact with the ground. This assumption implies that there is no active control between the end of the stance leg and the ground itself, which makes the system underactuated. The degree of underactuation in a single support is equal to one (five degrees of freedom and four actuators). The degree of underactuation during the flight phase is equal to three, with seven degrees of freedom and only four actuators [61]. Control design, motion planning, disturbance handling, and real-time experimental validation represent a key challenge in controlling this system [62,63] or similar types of underactuated bipedal robots [64,65].

Snake robots belong also to a category of biologically inspired system designed to replicate the characteristics of real snakes. They can be used on pipeline inspection, fire fighting [66], search and rescue, and image-guided surgery [67]. These robots are underactuated and are hyper-redundant due to their multiple degrees of freedom. The system is generally composed of $n$ connected serial identical links connected through $n-1$ joints. The first $n-1$ links are actuated, while the last $n$th link, referred to as the snake robot's head, remains passive. In this case, the orientation and position of the robot's center of mass are underactuated. Various studies in the literature have investigated the locomotion control problems for snake robots. For instance, optimal control [68] and virtual holonomic constraints [69] are proposed for snake robots without non-holonomic velocity constraints and computed torque control [70,71] for systems with non-holonomic constraints.

### 1.6.4 Underactuation in marine vehicles

The autonomous underwater vehicle (AUV) system depicted in Fig. 1.15 has three degrees of freedom $(x, y, \psi)$ that represent the positions in the $(x, y)$ plane and the orientation. It also has two control inputs, referred to as the thrusting force and the rudder. The first one is the force produced by the AUV's thrusters or actuators, used to enable the vehicle to move through the water, in a given direction. The second one is responsible for adjusting the rudder angle, the direction in which the AUV moves can be altered. This allows the vehicle to turn and maintain a specific trajectory. In this case, it is only possible to independently design two trajectories in the plane with three degrees of freedom. The stabilization point of the system is also referred to as stabilization control, a control input that guides the vehicle from an initial position to a specified target

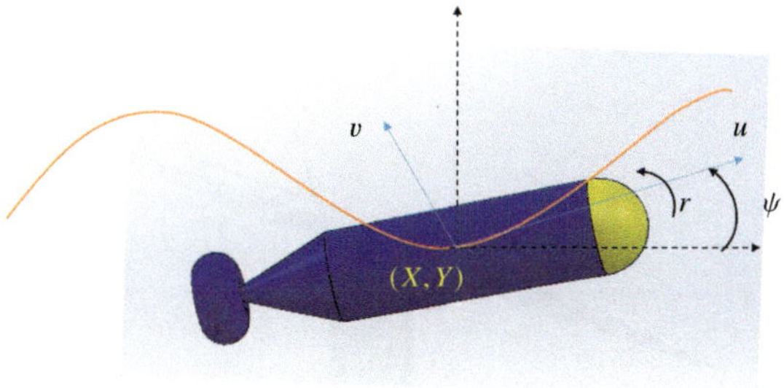


**FIGURE 1.15** Illustration of AUV model.

position and orientation, and stabilizes it in that position. Various interesting illustrative examples of underactuated AUV and control approaches can be found in [72–76]. Many factors contribute to the complexity of motion control of AUV such as the nonlinearity and coupling of the equation of motion, the time-varying parameters, and external disturbances.

### 1.6.5 Underactuated mechanical system for education purposes

The diversity and complexity of UMSs have led researchers in the field to analyze, on a case-by-case basis, examples of interesting low-dimensional underactuated mechanical systems (i.e., systems with few degrees of freedom). Despite being of low dimension, these systems, exhibit a non-zero degree of underactuation and highly nonlinear dynamics. Let us consider the example of an inverted pendulum on a cart illustrated in Fig. 1.1, a classic underactuated system commonly studied in control theory. The system consists of a cart that can move horizontally and a pendulum attached to the cart, which can rotate freely in the vertical plane. The problem of stabilization of the inverted pendulum in its unstable equilibrium position was addressed in [23]. However, the position of the cart is not regulated. A controlled Lagrangian method was presented and a stabilizing control law is obtained by identifying the closed-loop system. It is constructed by modifying the Lagrangian of the original system by modulating the kinetic energy of the system. The injection of a damping term stabilizes the system. In [24], the system is globally stabilized in its unstable equilibrium position. Indeed, the dynamics of the sub-systems, consisting of the different generalized coordinates, are decoupled by a first transformation, resulting in a cascade system. The size of this system is reduced by the backstepping technique. The resulting reduced system is then stabilized by a control law obtained by searching for the fixed point of the resulting reduced system. A method using flatness-based control was proposed in [25] to stabilize the inverted pendulum around its unstable equilibrium position. The system is first partially linearized. The resulting system is then approximated by a completely linear, flat system. The ball on the rail illustrated in Fig. 1.16 is also a low-order UMS. It consists

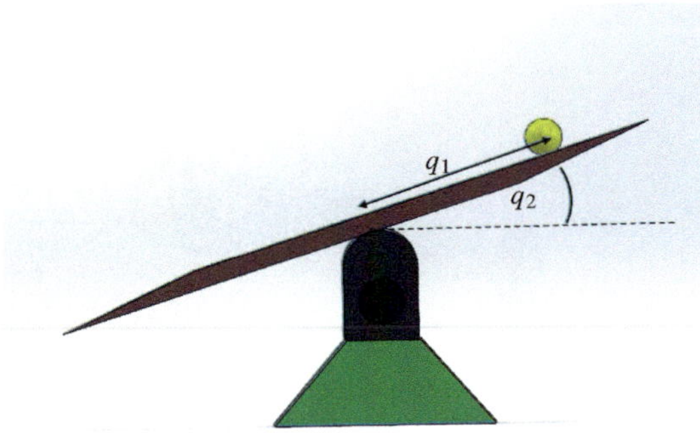


**FIGURE 1.16** The beam-and-ball system.

of a rail connected at its center to a base by an actuated pivot joint. This joint allows it to rotate in the vertical plane. The system also includes a ball that moves freely along this rail.

The authors in [26] addressed the problem of tracking reference trajectories on the ball only. A controller is proposed to linearize an approximation of this system, since the input-state feedback is not enough to completely linearize the system. This approximation is obtained by neglecting higher-order terms. Another backstepping control approach was proposed in [27] to reduce the order of the system. The resulting sub-system is then stabilized with a fixed-point controller.

The study proposed in [28] is based on a passivity control approach. It consists of extracting the kinetic and potential energy terms of the system separately, and then combining the obtained terms into a control law. The potential energy aims to make the equilibrium position attractive, and the kinetic energy stabilizes the system's dynamics by inserting damping terms.

In recent years, manipulation of objects by robots has also given rise to a growing interest in many applications, such as pushing boxes, pick and place, grasping operation, surgery tasks, etc.

The manipulation of objects can be classified as a prehensile or a non-prehensile task. As explained in [29], the human non-prehensile manipulation means that the object is not directly maintained between the fingertips or the palm of the hand. In contrast to an object retained by the palm of a human hand, the object in non-prehensile manipulation is subject to unilateral constraints and is not strictly constrained to follow the movement of the hand [30]. The problem of controlling such a manipulation is still far from being solved. It is a sophisticated dynamic manipulation that becomes an attractive topic for the robotic and automatic community given the difficulty of the motion planning and control. Indeed, the whole dynamics (i.e., the object and the manipulating hand) should be taken into account in the control design and the contact between the manipulator and the object should not be lost. Furthermore, these types of systems are often underactuated and sensitive to any external disturbances. Different

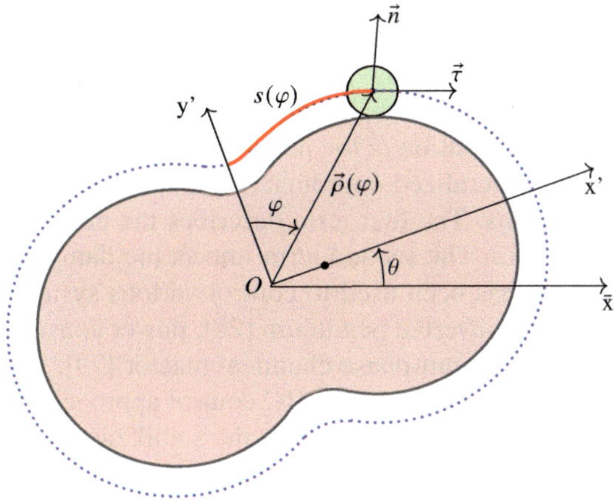


**FIGURE 1.17** General overview of the butterfly system.

low-degree-of-freedom systems illustrating a non-prehensile manipulation of a rolling movement, were designed and studied by the research community (i.e., the ball and beam [31], the disk on disk [30,32], and the butterfly robot [33,34], illustrated in Fig. 1.17). In this part, we are interested in the butterfly robot, as it presents a suitable example of a complex low-order UMS. The idea of designing such a system was first proposed in [35] to study a planar version of a juggler's skill with one joint robot. The system consists of two identical and parallel eight shaped plates with a gap smaller than the diameter of the ball. The rotational movement of these two platforms allows the rolling of a ball on the plates' boundaries. The dynamic model is designed in [33], which seems to be more accurate and precise than other reported approaches. Moreover, it considers the rotational energy of the ball and the reaction forces at the point of contact. A virtual-holonomic-constraints (VHC)-based motion planning [36] and transverse-linearization-based orbital stabilization [37] were proposed and implemented experimentally by [33]. An energy-based control was proposed by [34] to stabilize the robot at certain unstable equilibrium points. Another control approach was proposed in the literature for a rolling manipulation, based on the port-Hamiltonian transformation and a new passivity-based control approach [38].

## 1.7 Literature review about existing control strategies

Various control methods have been extensively studied in the literature and have been successfully applied to UMSs. In this section, we provide an overview of the control strategies proposed in the literature. It is worth noting that this is not a full review of all existing control approaches.

### 1.7.1 Passivity-based control

The interconnection and damping assignment passivity-based control (IDA-PBC) is introduced in [28] to globally stabilize the system around its unstable equilibrium (the upward position of the pendulum). The control law of the IDA-PBC depends on the generalized coordinates and momenta. It is described by the addition of two terms. The first term describes the energy shaping control to assign the equilibrium. The second term injects the damping to achieve the asymptotic stability. It has been used to control various systems, e.g., ball and beam [28], inertia wheel inverted pendulum [28], power converters [77], induction motor [78], non-minimum-phase chemical reactor [79], flexible spacecraft [80], among others. Based on the IDA-PBC control approach, a control scheme with bounded torque is designed in [81] for the stabilization of an UMS by including a tangent function in the control law. Unfortunately, this control scheme may demand a high input torque that makes a real-time implementation difficult to achieve. In [82], an extension of the standard IDA-PBC approach was applied to the Furuta pendulum. The goal was to bring the first link to a desired position equal to $-90°$ and the second to the upward vertical direction. Using Dahl's friction model and a nonlinear state observer, the friction at the actuated joint was compensated in the system. The IDA-PBC is a successful control approach in the context of mechanical systems described in Port-Controlled Hamiltonian (PCH) form. This latter has the advantage of covering a large set of physical systems and providing important structural properties. It invokes the principle of energy shaping and dissipation, where the closed-loop energy function may be obtained through the solution of partial differential equations (PDEs) called matching equations. The first drawback of such an approach is the difficulty of resolving the PDEs, especially for high-order UMSs. The transformation of the PDEs into ordinary differential equations (ODEs) is a method proposed in [83] with the aim of simplifying the resolution of the partial differential equations associated with the kinetic energy. The approach was based on the re-parametrization of the dynamics and transformation of the original system.

Another simplification of the PDEs associated with the potential energy was proposed in [84]. The main approach was based on a parametrization of the closed-loop inertia matrix. It has been applied to solve a global stabilization of a rotary inverted pendulum. The second shortcoming of the standard IDA-PBC approach is the need to enhance the robustness of the closed-loop system. Numerous solutions have been proposed in the literature. The inclusion of Integral Control (IC) with the standard IDA-PBC has been proposed in several works. For instance, in [85] a PID-like controller was proposed for the robustness problem of separable PCH systems. Assuming that the stabilization problem is solved by an IDA-PBC approach, the authors introduced an integral action to solve the robustness issue of a fully actuated system. This approach was extended to control underactuated mechanical systems where the IC was included on the non-passive output. The same idea was also proposed in [86] with validation by real-time experiments. It consists of adding an integral action on the non-passive outputs, to deal with the problem of constant and matched disturbance rejection in UMSs. Rejection of external disturbances in the IDA-PBC approach was also experimentally addressed in [87], by providing sufficient stability conditions to deal with matched and unmatched disturbances. However, when there are unmatched disturbances, the static error cannot be properly compensated by the controller, as stated by the authors. To address this drawback, a Model Reference Adaptive control approach (MRA-IDA-PBC) is proposed in [88,89] to compensate errors caused by uncertain parameters and unmatched disturbances. The approach has been successfully validated and implemented in a real underactuated device. A comparative experimental study of the aforementioned solutions was proposed in [90]. Based on the experimental results and the performance-evaluation indices, the PI-IDA-PBC showed significant performance in terms of convergence and energy consumption with respect to the adaptive controller.

### 1.7.2 Backstepping control

The standard backstepping is a recursive method based on construction of a control Lyapunov functions, especially for systems described in strict-feedback form. It is designed for the global stabilization of nonlinear systems and convergence of the system's variables to zero. It has been extensively applied to control underactuated mechanical systems.

In [91], the authors address the problem of global asymptotic stabilization and tracking of an autonomous underactuated surface vessel. After a global diffeomorphism change of coordinates, the dynamic model of the system is transformed into two interconnected sub-systems in cascade form. Then, the stabilization problem is reduced to control a third-order chained form by a discontinuous feedback law, derived based on the backstepping design approach.

A simplified dynamic model of an underactuated quadruped robot was developed in [92] considering the system as a planar five-link serial rigid-body chain model with six DOFs and four control inputs. Then, based on a coordinate transformation of the system into strict-feedback form, a robust backstepping controller is designed for the stabilization and tracking trajectory of the system.

In [93], the problem of stabilization of an underactuated system with a tree structure is addressed. Through a proposed change of coordinates, the tree structure can be transformed into a chain structure. As the system cannot be asymptotically stabilized by a single continuous feedback control law, the authors proposed a hybrid control law, switching between two control laws. The first one is designed using a backstepping procedure away from singularities and the second is a linear state feedback controller designed close to the singularities.

The main drawbacks of the backstepping approach lie in strictly transforming the exact mathematical model into the strict feedback form and the repeated differentiation of virtual inputs. This may induce an explosion of undefinable terms, especially for high-order models. Moreover, the exact mathematical equations of a real nonlinear system are difficult to obtain, as variable uncertainties are abundant in nonlinear UMSs and may depend on both time and states.

### 1.7.3 Model predictive control

Predictive control has emerged as one of the most used control algorithms for nonlinear systems. The main idea behind predictive control is to predict the behavior of the system based on its dynamic model and to solve an optimization problem at each time step to find the optimal control action that drives the predicted plant output to the desired reference trajectory [94]. There are two main types of model predictive controllers (MPCs), a linear and nonlinear MPC. The first one is based on a linear model to predict the nonlinear system dynamics at a specific operating point, while the nonlinear MPC (NMPC) approach is based on a nonlinear model. Due to its linearity, the computational time of the linear MPC is less than that of the NMPC. However, if the nonlinearity of the system is too high, the linear MPC is not recommended to be used, as linear models are often inadequate to describe the dynamics. The difference between both classes is that the NMPC requires a longer running time, which is computationally expensive and not recommended, especially for fast systems.

A comparative study between two MPC approaches is proposed in [95]. The first proposed control strategy is based on a generalized predictive control (GPC) [96] and the second one is an extension of GPC based on Laguerre functions, used to reduce the optimization problem. Both controllers are proposed to solve the stabilization problem of the inertia wheel inverted pendulum.

Another approach based on GPC was also proposed by [97] to deal with the stabilization problem of the underactuated IWIP. The performance of the proposed control scheme and its robustness towards uncertainties and external disturbances are tested through real-time experiments. A model predictive stabilization problem of a spacecraft with two bounded control torques was proposed in [98]. The authors propose a systematic approach to design the parameters for the MPC algorithm, satisfying the asymptotic stability of the closed-loop system.

### 1.7.4 Sliding mode control

Sliding mode control (SMC) has received increased attention within the control research community due to (i) its robustness against external disturbances (ii) its insensitivity to parameter variation and (iii) its wide range of applications. The design of the SMC approach consists mainly of two fundamental steps. The first one is to ensure the reachability phase, and the second is to drive the states of the system to the equilibrium point. In the reachability phase, the controller drives the states to a stable domain called the sliding surface. Then, during the sliding phase, the controller keeps the states on the surface while sliding to the equilibrium point.

First- and second-order SMCs are widely reported in the literature to control UMSs, such as a surface vessel [99], helicopter [100], ball- and beam system [101], pendubot [102], TORA system [103], and the inertia wheel inverted pendulum [104,105]. The quasi-homogeneous control belongs to the set of second-order sliding mode controller and has apparently some advantages and restriction of the first-order sliding mode controller. Indeed, it is characterized by its robustness with respect to the disturbances and finite-time convergence on a predefined sliding surface. However, the main drawback of this approach is the chattering phenomenon. In [106] the quasi-homogeneous control is used to stabilize locally the inverted pendulum around its unstable equilibrium position while forcing the cart to converge to a desired end position. The desired stability property of the closed-loop system is provided, driving the system to the zero dynamics manifold in a finite time.

In [107], multiple sliding surfaces control based on a partial feedback linearization is proposed using a change of coordinates that transforms the system into two cascaded sub-systems. The implicit control technique is employed to design the controller of the inner sub-system that has a virtual input appearing from the linear system in a non-affine manner. This sub-system is called by the author the non-affine nonlinear core sub-system. A high-order sliding mode (HOSM) control approach, namely quasi-continuous, is designed by [108]. It is a high-order sliding mode structure composed of a controller and differentiator that allows stabilizing the internal dynamics of the inertia wheel inverted pendulum and to track accurately the predefined sine shape trajectory. However, the well-known limitation of the SMC is the chattering phenomenon induced by the control law's discontinuous term. It may cause dangerous damage to the actuators and affect control accuracy, especially in real-time implementation. To overcome the chattering problem, some known solutions were proposed in the literature, the continuous approximation of the discontinuous control law using a saturation function [39], the observer-based approach [109] or the design of second-order sliding mode controllers [110,111]. Other effective control solutions are designed in the literature based on fuzzy or neural algorithms [112] [113]. However, they require a careful selection of parameters and considerable computing time, especially in real-time experiments. It is well demonstrated that a continuous second-order sliding mode controller can attenuate the chattering, although second-order SMCs require a priori knowledge of the uncertainties boundaries that are difficult to practically estimate in real time for nonlinear UMSs.

### 1.7.5 Intelligent controllers

The model-free control introduced by [114] has numerous successful applications in various areas. The main advantage of this controller (also called Intelligent-PID) is that it does not require prior knowledge of the mathematical model, nor complex tuning of the parameters. Accordingly, it can easily be applied to control systems with unknown dynamics. A dual model-free controller, proposed by [46], performs the tracking of a periodic, continuous, and derivable trajectory of non-actuated coordinates of the pendubot and cart-pole system. The proposed approach is designed to stabilize the internal dynamics of the closed-loop system, and to generate a stable limit cycle on both coordinates. It chooses the right trajectory parameter in order to stabilize the actuated coordinate. The designed controller scheme inherits the advantages of model-free control because neither mathematical modeling nor complex parameter identification are needed.

Fuzzy control belongs to the set of intelligent controllers. It is an effective approach to solve the problem of uncertainties in control systems and does not require a mathematical model. It can also work with imprecise inputs, can handle nonlinearity, and is known for its insensitivity to disturbances. Fuzzy control also has some drawbacks. The control precision of fuzzy control is low. The membership function and control rules of the fuzzy controller are presummarized based on experience. It can not be corrected during the control process and does not have the ability to learn and adapt. A survey of the application control methods based on fuzzy systems for underactuated systems during the years 2000 to 2017 is presented in [115].

In [116], the authors propose a new fuzzy logic control based on two different fuzzy reasoning control laws for the swinging up and balancing of the pendubot.

Other hybrid control approaches proposed in the literature were based on Fuzzy Sliding Mode Control (FSMC) [117], Adaptive Fuzzy Sliding Mode Control (AFSMC) [112], and Neural Fuzzy Control (NFC) [118].

## 1.8 Conclusion

This chapter presents the problematic challenges associated with different classes of mechanical systems. We then demonstrated some illustrations of underactuated systems and explained different control problems associated with UMSs. Finally, we presented an overview about control approaches that can be used to deal with the stabilization and stable limit cycles. This chapter may be considered as a summary about stabilization and control challenges of nonlinear UMSs that have seen significant progress through classical and hybrid approaches.

## References

[1] N. Lobontiu, System Dynamics for Engineering Students: Concepts and Applications, second edition, Academic Press, Boston, 2018.

[2] A. Müller, T. Hufnagel, A projection method for the elimination of contradicting control forces in redundantly actuated PKM, in: Proceedings of the 2011 IEEE International Conference on Robotics and Automation, Shanghai, China, 2011, pp. 3218–3223.

[3] M. Bennehar, Some contributions to nonlinear adaptive control of PKMs: from design to real time experiments, PhD thesis, University of Montpellier, 2015.

[4] M. Luces, J.K. Mills, B. Benhabib, A review of redundant parallel kinematic mechanisms, Journal of Intelligent & Robotic Systems 86 (2017) 175–198.
[5] W. Zhang, M. Brunner, L. Ott, M. Kamel, R. Siegwart, J. Nieto, Learning dynamics for improving control of overactuated flying systems, IEEE Robotics and Automation Letters 5 (4) (2020) 5283–5290.
[6] Y. Michellod, P. Mullhaupt, D. Gillet, On achieving periodic joint motion for redundant robots, in: IFAC Proceedings Volumes, 17th IFAC World Congress, vol. 41(2), 2008, pp. 4355–4360.
[7] B.J. Emran, H. Najjaran, A review of quadrotor: an underactuated mechanical system, Annual Reviews in Control 46 (2018) 165–180, https://doi.org/10.1016/j.arcontrol.2018.10.009.
[8] B. He, S. Wang, Y. Liu, Underactuated robotics: a review, International Journal of Advanced Robotic Systems 16 (4) (2019) 1–29, https://doi.org/10.1177/1729881419862164.
[9] A. Hfaiedh, A. Chemori, A. Abdelkrim, Observer-based robust integral of the sign of the error control of class I of underactuated mechanical systems: theory and real-time experiments, Transactions of the Institute of Measurement and Control 44 (2) (2022) 339–352.
[10] S. Krafes, Z. Chalh, A. Saka, A review on the control of second order underactuated mechanical systems, IEEE Transactions on Automatic Control (2018) (2018) 1–17, https://doi.org/10.1155/2018/9573514.
[11] M.W. Spong, The swing up control problem for the acrobot, IEEE Control Systems Magazine 15 (1) (1995) 49–55, https://doi.org/10.1109/37.341864.
[12] M. Zhang, T-J. Tarn, Hybrid control for the pendubot, in: Proceedings 2001 ICRA, IEEE International Conference on Robotics and Automation, vol. 2, 2001, pp. 2102–2107.
[13] A. Choukchou-Braham, B. Cherki, M. Djemaï, K. Busawon, Classification of underactuated mechanical systems, in: Analysis and Control of Underactuated Mechanical Systems, Springer, Cham, 2014.
[14] D. Seto, J. Baillieul, Control problems in super-articulated mechanical systems, IEEE Transactions on Automatic Control 39 (12) (1994) 2442–2453.
[15] R. Olfati-Saber, Nonlinear control of underactuated mechanical systems with application to robotics and aerospace vehicles, PhD thesis, Massachusetts Institute of Technology, 2001.
[16] C.Y. Su, Y. Stepanenko, Robust motion/force control of mechanical systems with classical nonholonomic constraints, IEEE Transactions on Automatic Control 39 (3) (1994) 609–614.
[17] H. Arai, K. Tanie, N. Shiroma, Nonholonomic control of a three-dof planar underactuated manipulator, IEEE Transactions on Robotics and Automation 14 (5) (1998) 681–695.
[18] E. Jarzebowska, Stabilizability and motion tracking conditions for mechanical nonholonomic control systems, Mathematical Problems in Engineering 2007 (2007) 20, https://doi.org/10.1155/2007/31267.
[19] Y. Liu, H. Yu, A survey of underactuated mechanical systems, IET Control Theory & Applications 7 (2013) 921–935, https://doi.org/10.1049/iet-cta.2012.0505.
[20] M. Yue, P. Hu, W. Sun, Path following of a class of non-holonomic mobile robot with underactuated vehicle body, IET Control Theory & Applications 4 (10) (2010) 1898–1904.
[21] Y. Tian, N. Sidek, N. Sarkar, Modeling and control of a nonholonomic wheeled mobile robot with wheel slip dynamics, in: 2009 IEEE Symposium on Computational Intelligence in Control and Automation, 2009, pp. 7–14.
[22] R.W. Brockett, Asymptotic stability and feedback stabilization, in: Differential Geometric Control Theory, Birkhäuser, 1983, pp. 181–191.
[23] A.M. Bloch, N.E. Leonard, J.E. Marsden, Stabilization of the pendulum on a rotor arm by the method of controlled Lagrangians, in: Proceedings of the International Conference on Robotics and Automation, 1999, pp. 500–505.
[24] R. Olfati-Saber, Fixed point controllers and stabilization of the cart-pole system and the rotating pendulum, in: Proceedings of the 38th Conference on Decision and Control, 1999, pp. 1174–1181.
[25] C. Aguilar-Ibáñez, H. Sira-Ramírez, M.S. Suárez-Castañón, A flatness-based controller for the stabilization of the inverted pendulum, Mathematical Problems in Engineering 2011 (1) (2011) 659685, https://doi.org/10.1155/2011/659685.

[26] J. Hauser, S. Sastry, P. Kokotović, Nonlinear control via approximate input-output linearization: the ball and beam example, IEEE Transactions on Automatic Control 37 (1992) 392–398.
[27] R. Olfati-Saber, A. Megretski, Controller design for the beam-and-ball system, in: Proceedings of the 37th Conference on Decision and Control, vol. 4, 1998, p. 4555.
[28] R. Ortega, et al., Stabilization of underactuated mechanical systems via interconnection and damping assignment, IEEE Transactions on Automatic Control 47 (2000) 1218–1233.
[29] D. Serra, Robot control for nonprehensile dynamic manipulation tasks, in: International Conference on Informatics in Control, Automation and Robotics, 2016.
[30] D. Serra, F. Ruggiero, A. Donaire, L.R. Buonocore, V. Lippiello, B. Siciliano, Control of nonprehensile planar rolling manipulation: a passivity-based approach, IEEE Transactions on Robotics 35 (2) (2019) 317–329.
[31] F. Gordillo, J. Aracil, F. Gomez-Estern, Stabilization of autonomous oscillations and the Hopf bifurcation in the ball and beam, in: Proceedings of the 41st IEEE Conference on Decision and Control, Las Vegas, NV, USA, 2002, pp. 3924–3925.
[32] J.C. Ryu, F. Ruggiero, K.M. Lynch, Control of nonprehensile rolling manipulation: balancing a disk on a disk, IEEE Transactions on Robotics 29 (5) (2013) 1152–1161, https://doi.org/10.1109/TRO.2013.2262775.
[33] M. Surov, A. Shiriaev, L. Freidovich, S. Gusev, L. Paramonov, Case study in non-prehensile manipulation: planning and orbital stabilization of one-directional rollings for the "butterfly" robot, in: 2015 IEEE International Conference on Robotics and Automation (ICRA), 2015, pp. 1484–1489.
[34] M. Cefalo, L. Lanari, G. Oriolo, Energy-based control of the butterfly robot, in: IFAC Proceedings Volumes, vol. 39, 2006, pp. 1–6.
[35] K.M. Lynch, N. Shiroma, H. Arai, K. Tanie, The roles of shape and motion in dynamic manipulation: the butterfly example, in: Proceedings of the 1998 IEEE International Conference on Robotics and Automation (Cat. No. 98CH36146), 1998, pp. 1958–1963.
[36] M. Maggiore, L. Consolini, Virtual holonomic constraints for Euler–Lagrange systems, IEEE Transactions on Automatic Control 58 (4) (2013) 1001–1008, https://doi.org/10.1109/TAC.2012.2215538.
[37] A.S. Shiriaev, L.B. Freidovich, S.V. Gusev, Transverse linearization for controlled mechanical systems with several passive degrees of freedom, IEEE Transactions on Automatic Control 55 (4) (2010) 893–906, https://doi.org/10.1109/TAC.2010.2042000.
[38] A. Donaire, F. Ruggiero, L.R. Buonocore, V. Lippiello, B. Siciliano, Passivity-based control for a rolling-balancing system: the nonprehensile disk-on-disk, IEEE Transactions on Control Systems Technology 25 (6) (2017) 2135–2142, https://doi.org/10.1109/TCST.2016.2637719.
[39] J.J. Slotine, W. Li, Applied Nonlinear Control, Prentice-Hall, ISBN 0130400491, 1991.
[40] Y. Hu, G. Yan, Z. Lin, Stable running of a planar underactuated biped robot, Robotica 29 (5) (2011) 657–665, https://doi.org/10.1017/S0263574710000512.
[41] A.S. Shiriaev, J.W. Perram, C. Canudas-de Wit, Constructive tool for orbital stabilization of underactuated nonlinear systems: virtual constraints approach, IEEE Transactions on Automatic Control 50 (8) (2005) 1164–1176, https://doi.org/10.1109/TAC.2005.852568.
[42] L.B. Freidovich, P. La Hera, U. Mettin, A. Robertsson, A.S. Shiriaev, R. Johansson, Shaping stable periodic motions of inertia wheel pendulum: theory and experiment, Asian Journal of Control 11 (5) (2009) 548–556, https://doi.org/10.1002/asjc.135, https://onlinelibrary.wiley.com/doi/abs/10.1002/asjc.135.
[43] D.M. Alonso, E.E. Paolini, J.L. Moiola, Global bifurcation analysis of a controlled underactuated mechanical system, Nonlinear Dynamics 40 (3) (2005) 205–225, https://doi.org/10.1007/s11071-005-6188-z.
[44] H. Gritli, N. Khraief, A. Chemori, S. Belghith, Self-generated limit cycle tracking of the underactuated inertia wheel inverted pendulum under IDA-PBC, Nonlinear Dynamics 89 (3) (2017) 2195–2226, https://doi.org/10.1007/s11071-017-3578-y.

[45] S. Andary, A. Chemori, S. Krut, Estimation-based disturbance rejection in control for limit cycle generation on inertia wheel inverted pendulum testbed, in: 2009 IEEE/RSJ International Conference on Intelligent Robots and Systems, St. Louis, MO, USA, Oct. 2009, pp. 1302–1307.
[46] S. Andary, A. Chemori, A dual model-free control of non-minimum phase systems for generation of stable limit cycles, in: 2011 50th IEEE Conference on Decision and Control and European Control Conference, Orlando, FL, USA, 2011, pp. 1387–1392.
[47] J. Hauser, S. Sastry, G. Meyer, Nonlinear control design for slightly non-minimum phase systems: Application to V/STOL aircraft, Automatica 28 (4) (1992) 665–679.
[48] Z. Li, A simple nonlinear stabilization control of planar vertical take-off and landing aircraft based on finite-time control and controlled Lagrangians, Asian Journal of Control 24 (4) (2022) 1862–1872.
[49] S.P. Liu, Z. Lu, L.C. Wu, Nonlinear trajectory tracking control of under-actuated VTOL aircraft, Xitong Fangzhen Xuebao / Journal of System Simulation 18 (2) (2006) 735–737.
[50] M. Heidarian, A.Y. Memon, Attitude control of VTOL-UAVs, in: Proceedings of the 2012 UKACC International Conference on Control, Cardiff, 2012, pp. 363–368.
[51] T. McGeer, Passive dynamic walking, The International Journal of Robotics Research 9 (2) (1990) 62–82, https://doi.org/10.1177/027836499000900206.
[52] M. Garcia, A. Chatterjee, A. Ruina, M. Coleman, The simplest walking model: stability, complexity, and scaling, Journal of Biomechanical Engineering 120 (4) (1998) 281–288.
[53] J. Moon, M. Spong, Bifurcations and chaos in passive walking of a compass-gait biped with asymmetries, in: Proceedings – IEEE International Conference on Robotics and Automation, 2010, pp. 1721–1726.
[54] H. Gritli, Occasional stabilisation of limit cycle walking and control of chaos in the passive dynamics of the compass-gait biped model, International Journal of Automation and Control 15 (3) (2021) 198–225.
[55] H. Gritli, N. Khraief, S. Belghith, Complex walking behaviours, chaos and bifurcations of a simple passive compass-gait biped model suffering from leg length asymmetry, International Journal Of Simulation And Process Modelling 13 (2018) 446–462.
[56] A. De Luca, B. Siciliano, Trajectory control of a non-linear one-link flexible arm, International Journal of Control 50 (5) (1989) 1699–1715.
[57] G. Wang, Simulations of the trajectory tracking and vibration suppression control of flexible-link robot arm, in: Proceedings of the 2010 International Conference on Intelligent System Design and Engineering Application, Changsha, China, 2010, pp. 387–390.
[58] M. Pierallini, F. Angelini, R. Mengacci, A. Palleschi, A. Bicchi, M. Garabini, Trajectory tracking of a one-link flexible arm via iterative learning control, in: Proceedings of the 2020 IEEE/RSJ International Conference on Intelligent Robots and Systems (IROS), Las Vegas, NV, USA, 2020, pp. 7579–7586.
[59] D.A. Bristow, M. Tharayil, A.G. Alleyne, A survey of iterative learning control, IEEE Control Systems Magazine 26 (3) (2006) 96–114.
[60] E.R. Westervelt, G. Buche, J.W. Grizzle, Inducing dynamically stable walking in an underactuated prototype planar biped, in: Proceedings – IEEE International Conference on Robotics and Automation, 2004, pp. 4234–4239.
[61] C. Chevallereau, et al., RABBIT: a testbed for advanced control theory, IEEE Control Systems Magazine 23 (5) (2003) 57–79.
[62] C. Sabourin, O. Bruneau, J-G. Fontaine, Start, stop and transition of velocities of an underactuated bipedal robot without reference trajectories, International Journal of Humanoid Robotics 1 (2) (2004) 349–374.
[63] C. Sabourin, O. Bruneau, G. Buche, Experimental validation of a robust control strategy for the robot RABBIT, in: Proceedings – IEEE International Conference on Robotics and Automation, 2005, pp. 2393–2398.
[64] Y. Gao, Y. Gong, V. Paredes, A. Hereid, Gu Y. Time-varying, ALIP model and robust foot-placement control for underactuated bipedal robotic walking on a swaying rigid surface, in: 2023 American Control Conference (ACC), 2023, pp. 3282–3287.

[65] S. Gupta, A. Kumar, A brief review of dynamics and control of underactuated biped robots, Advanced Robotics 31 (12) (2017) 607–623.

[66] P. Liljeback, O. Stavdahl, A. Beitnes, SnakeFighter – development of a water hydraulic fire fighting snake robot, in: 2006 9th International Conference on Control, Automation, Robotics and Vision, Singapore, 2006, pp. 1–6.

[67] S. Tully, G. Kantor, M.A. Zenati, H. Choset, Shape estimation for image-guided surgery with a highly articulated snake robot, in: 2011 IEEE/RSJ International Conference on Intelligent Robots and Systems, San Francisco, CA, USA, 2011, pp. 1353–1358.

[68] G. Hicks, K. Ito, A method for determination of optimal gaits with application to a snake-like serial-link structure, IEEE Transactions on Automatic Control 50 (9) (2005) 1291–1306.

[69] E. Rezapour, K.Y. Pettersen, P. Liljebäck, et al., Path following control of planar snake robots using virtual holonomic constraints: theory and experiments, Robotics and Biomimetics 1 (2014) 3.

[70] F. Matsuno, H. Sato, Trajectory tracking control of snake robots based on dynamic model, in: Proceedings of the 2005 IEEE International Conference on Robotics and Automation, Barcelona, Spain, 2005, pp. 3029–3034.

[71] H. Date, Y. Hoshi, M. Sampei, Locomotion control of a snake-like robot based on dynamic manipulability, in: Proceedings. 2000 IEEE/RSJ International Conference on Intelligent Robots and Systems (IROS 2000), Takamatsu, Japan, vol. 3, 2000, pp. 2236–2241.

[72] J-H. Li, P-M. Lee, Path tracking in dive plane for a class of torpedo-type underactuated AUVs, in: Proceedings of the 2009 7th Asian Control Conference, Hong Kong, China, 2009, pp. 360–365.

[73] G.R. Cho, D-G. Park, H. Kang, M-J. Lee, J-H. Li, Horizontal trajectory tracking of underactuated AUV using backstepping approach, IFAC-PapersOnLine 52 (16) (2019) 174–179.

[74] Z. Zheng, L. Ruan, M. Zhu, Output-constrained tracking control of an underactuated autonomous underwater vehicle with uncertainties, Ocean Engineering 175 (2019) 241–250.

[75] Q. Guo, Y. Li, H. Liu, J. Xiang, An underactuated AUV tracking algorithm based on backstepping adaptive sliding mode control, in: Proceedings of the 2020 Chinese Automation Congress (CAC), Shanghai, China, 2020, pp. 3765–3770.

[76] J. Li, S. Li, Underactuated autonomous underwater vehicle trajectory tracking control in three dimensions based on fractional order sliding mode, in: Proceedings of the 2023 IEEE International Conference on Mechatronics and Automation (ICMA), Harbin, Heilongjiang, China, 2023, pp. 1340–1345.

[77] H. Rodriguez, R. Ortega, G. Escobar, N. Barabanov, A robustly stable output feedback saturated controller for the boost dc-to-dc converter, Systems & Control Letters 40 (1) (2000) 1–8, https://doi.org/10.1016/S0167-6911(99)00113-9.

[78] H. Gonzalez, M. Duarte-Mermoud, I. Pelissier, J. Travieso, R. Ortega, A novel induction motor control scheme using IDA-PBC, Journal of Control Theory and Applications 6 (2008) 59–68, https://doi.org/10.1007/s11768-008-7193-9.

[79] H. Ramírez, D. Sbarbaro, R. Ortega, On the control of non-linear processes: an IDA-PBC approach, Journal of Process Control 19 (3) (2009) 405–414, https://doi.org/10.1016/j.jprocont.2008.06.018.

[80] S. Aoues, D. Matignon, D. Alazard, Control of a flexible spacecraft using discrete IDA-PBC design, in: 5th IFAC Workshop on Lagrangian and Hamiltonian Methods for Nonlinear Control (LHMNC2015), Lyon, France, 2015, pp. 188–193.

[81] V. Santibanez, R. Kelly, J. Sandoval, Control of the inertia wheel pendulum by bounded torques, in: Proceedings of the 44th IEEE Conference on Decision and Control, 2005, pp. 8266–8270.

[82] J. Sandoval, R. Kelly, V. Santibáñez, Interconnection and damping assignment passivity-based control of a class of underactuated mechanical systems with dynamic friction, International Journal of Robust and Nonlinear Control 21 (2011) 738–751, https://doi.org/10.1002/rnc.1622.

[83] G. Viola, R. Ortega, R. Banavar, J.A. Acosta, Astolfi A. Total, Energy shaping control of mechanical systems: simplifying the matching equations via coordinate changes, IEEE Transactions on Automatic Control 52 (6) (2007) 1093–1099, https://doi.org/10.1109/TAC.2007.899064.

[84] M. Ryalat, D. Laila, IDA-PBC for a class of underactuated mechanical systems with application to a rotary inverted pendulum, in: 52nd IEEE Conference on Decision and Control, Firenze, Italy, 2013, pp. 5240–5245.

[85] M. Ryalat, D.S. Laila, M.M. Torbati, Integral IDA-PBC and PID-like control for port-controlled Hamiltonian systems, in: 2015 American Control Conference (ACC), Chicago, United States, 2015, pp. 5365–5370.

[86] A. Donaire, J.G. Romero, R. Ortega, B. Siciliano, M. Crespo, Robust IDA-PBC for underactuated mechanical systems subject to matched disturbances, International Journal of Robust and Nonlinear Control 27 (2017) 1000–1016.

[87] N. Khraief-Haddad, A. Chemori, S. Belghith, External disturbance rejection in IDA-PBC controller for underactuated mechanical systems: From theory to real time experiments, in: 2014 IEEE Conference On Control Applications (CCA), 2014, pp. 1747–1752.

[88] N. Khraief-Haddad, A. Chemori, J. Pena, S. Belghith, Stabilization of inertia wheel inverted pendulum by model reference adaptive IDA-PBC: from simulation to real-time experiments, in: 2015 3rd International Conference on Control, Engineering Information Technology (CEIT), 2015, pp. 1–6.

[89] N. Khraief-Haddad, A. Chemori, S. Belghith, Robustness enhancement of IDA-PBC controller in stabilising the inertia wheel inverted pendulum: theory and real-time experiments, International Journal of Control 91 (2018) 2657–2672.

[90] A. Hfaiedh, A. Chemori, A. Abdelkrim, Stabilization of the inertia wheel inverted pendulum by advanced IDA-PBC based controllers: comparative study and real-time experiments, in: 17th International Multi-Conference on Systems, Signals and Devices (SSD'20), 2020, pp. 757–764.

[91] J. Ghommam, F. Mnif, A. Benali, N. Derbel, Asymptotic backstepping stabilization of an underactuated surface vessel, IEEE Transactions on Control Systems Technology 14 (2006) 1150–1157.

[92] H. Kazemi, V. Majd, M.M. Moghaddam, Modeling and robust backstepping control of an underactuated quadruped robot in bounding motion, Robotica 31 (2013) 423–439.

[93] A. Choukchou-Braham, B. Cherki, M. Djemaï, A backstepping procedure for a class of underactuated systems with tree structure, in: 2011 International Conference on Communications, Computing and Control Applications (CCCA), 2011, pp. 1–6.

[94] C. García, D. Prett, M. Morari, Model predictive control: theory and practice—a survey, Automatica 25 (1989) 335–348.

[95] M. Ahsan, M. Khalid, O. Kamal, Stabilization of an inertia wheel inverted pendulum using model-based predictive control, in: 2017 14th International Bhurban Conference on Applied Sciences and Technology (IBCAST), 2017, pp. 265–270.

[96] D. Clarke, C. Mohtadi, P. Tuffs, Generalized predictive control—Part I. The basic algorithm, Automatica 23 (1987) 137–148.

[97] N. Touati, A. Chemori, Predictive control for the stabilization of a fast mechatronic system: from simulation to real-time experiments, IFAC Proceedings Volumes 46 (5) (2013) 237–242.

[98] H. Li, W. Yan, Y. Shi, Continuous-time model predictive control of under-actuated spacecraft with bounded control torques, Automatica 75 (2017) 144–153.

[99] H. Ashrafiuon, K. Muske, L. McNinch, R. Soltan, Sliding-mode tracking control of surface vessels, IEEE Transactions on Industrial Electronics 55 (2008) 4004–4012.

[100] H. Rios, A. Rosales, A. Ferreira, A. Davilay, Robust regulation for a 3-DOF helicopter via sliding-modes control and observation techniques, in: Proceedings of the 2010 American Control Conference, 2010, pp. 4427–4432.

[101] R. Hirschorn, Incremental sliding mode control of the ball and beam, IEEE Transactions on Automatic Control 47 (2002) 1696–1700.
[102] D. Qian, J. Yi, D. Zhao, Hierarchical sliding mode control to swing up a pendubot, in: Proceedings of the 2007 American Control Conference, New York, NY, 2007, pp. 5254–5259.
[103] T. Wu, W. Gui, D. Hu, C. Du, Adaptive fuzzy sliding mode control for translational oscillator with rotating actuator: a fuzzy model, IEEE Access 6 (2018) 55861–55869.
[104] A. Hfaiedh, A. Abdelkrim, Revisited adaptive sliding mode control of underactuated mechanical systems with real-time experiments, International Journal of Modelling, Identification and Control 37 (3) (2021) 344–353.
[105] A. Hfaiedh, A. Chemori, A. Abdelkrim, Disturbance observer-based super-twisting control for the inertia wheel inverted pendulum, in: 2020 17th International Multi-Conference on Systems, Signals & Devices (SSD), Monastir, Tunisia, 2020, pp. 747–752.
[106] S. Riachy, Y. Orlov, T. Floquet, R. Santiesteban, J.P. Richard, Second-order sliding mode control of underactuated mechanical systems I: local stabilization with application to an inverted pendulum, International Journal of Robust and Nonlinear Control 18 (2008) 529–543.
[107] N. Qaiser, S. Tariq, N. Haq, T. Aziz, A novel nonlinear implicit sliding surface controller design for inertia wheel pendulum, in: Proceedings of the 10th WSEAS International Conference on Robotics, Control and Manufacturing Technology, Corfu, Greece, 2010, pp. 21–25.
[108] C. Zayane-Aissa, T. Laleg-Kirati, A. Chemori, Control of a perturbed under-actuated mechanical system, in: Proceedings of the 2015 IEEE Conference on Control Applications (CCA), Sydney, Australia, 2015, pp. 294–299.
[109] A. Bondarev, S. Bondarev, N. Kostyleva, V. Utkin, Sliding modes in systems with asymptotic state observers, Avtomatika I Telemehanika (Automation and Remote Control) 46 (1985) 679–684.
[110] G. Bartolini, A. Ferrara, E. Usai, Chattering avoidance by second-order sliding mode control, IEEE Transactions on Automatic Control 43 (2) (1998) 241–246.
[111] T. Gonzalez, J. Moreno, L. Fridman, Variable gain super-twisting sliding mode control, IEEE Transactions on Automatic Control 57 (8) (2012) 2100–2105.
[112] A. Markazi, M. Maadani, S. Zabihifar, N. Doost-Mohammadi, Adaptive fuzzy sliding mode control of under-actuated nonlinear systems, International Journal of Automation and Computing 15 (2018) 364–376.
[113] S. Mahjoub, F. Mnif, N. Derbel, Radial-basis-functions neural network sliding mode control for underactuated manipulators, in: Proceedings of the 10th International Multi-Conferences on Systems, Signals & Devices (SSD13), Hammamet, Tunisia, 2013, pp. 1–6.
[114] M. Fliess, C. Join, Model-free control and intelligent PID controllers: towards a possible trivialization of nonlinear control?, IFAC Proceedings Volumes, 15th IFAC Symposium on System Identification 42 (10) (2009) 1531–1550.
[115] X. Huang, A.L. Ralescu, H. Gao, H. Huang, A survey on the application of fuzzy systems for underactuated systems, Proceedings of the Institution of Mechanical Engineers, Part I, Journal of Systems and Control Engineering 233 (3) (2019) 217–244.
[116] X.Q. Ma, C.Y. Su, Theory and implementation of a fuzzy control scheme for pendubot, IFAC Proceedings Volumes 35 (1) (2002) 335–340.
[117] Y. Chang, C. Chang, C. Tao, H. Lin, J. Taur, Fuzzy sliding-mode control for ball and beam system with fuzzy ant colony optimization, Expert Systems with Applications 39 (2012) 3624–3633.
[118] A. Kharola, P.P. Patil, Neural fuzzy control of ball and beam system, International Journal of Energy Optimization and Engineering 6 (2) (2017) 64–78.

Chapter 2

# The inertia wheel inverted pendulum case study

**Contents**



## 2.1 Introduction

Low-order mechanical systems are generally characterized by a limited number of state variables. They can serve as a case study for early-stage analysis, simulation, and control design. They are often less costly and can provide a clear illustration of the behavior of more complex high-order systems. This is because they can highlight some key features and behaviors of the system, which can be obscured in the complexity of the high-order system. If we consider the example of the inverted pendulum, it has one degree of freedom and it represents a classic problem in dynamics and control theory. However, it illustrates various engineering challenges [1]. For example, human walking, a highly complex biomechanical process involving multiple joints and muscles, can be approximated using a model of two coupled inverted pendulums. In this simplified representation, each leg acts as an inverted pendulum during the walking motion [2]. During the single support phase, when a biped robot is supporting its body on one leg, the system can be represented by a three-dimensional single inverted pendulum model [3]. An interesting and clear review of different pendulum structures used in engineering applications is given in [4]. This can significantly help in better understanding the tackled problem as well as the proposed solution. In this chapter, we will present the description of a well-known


https://doi.org/10.1016/B978-0-44-324020-1.00009-9

system in the literature, called the inertia wheel inverted pendulum (IWIP). The concept of the IWIP is somewhat old; however, it is still an interesting typical benchmark. For instance, the same behavior of an inertia wheel inverted pendulum can be considered as a stabilizer for self-balancing electric motorcycles. It serves also as a simplified model of a monorail tram [5]. Underactuation in this system leads to an unstable and highly nonlinear dynamic behavior [12,13]. IWIP is a significant system in control theory due to its complex, nonlinear, and underactuated nature. It serves as an excellent testbed to test, analyze, validate, and compare various control strategies. Due to the reasons mentioned above, this low-order system is commonly utilized in the literature and has attracted considerable interest from researchers, such as in [6–12,44,45]. In the following sections, we will provide a detailed description of the system.

## 2.2 System's detailed description

The Inertia Wheel Inverted pendulum is a nonlinear reduced-order example of an underactuated mechanical system. It inherits the same features of much more complex systems (e.g., human balance support devices, gyrostabilizers, and stabilizers for self-balancing electric motorcycles). It is characterized by two degrees of freedom and one control input. The system includes an unactuated joint between the frame and pendulum body and an actuated joint between the body and inertia wheel, as illustrated in Fig. 2.1. Both revolute joints operate in the same plane.

In the simplified diagram of the system, we can consider that the inertia wheel (IW) is equivalent to two identical point masses located at the action points of both forces $F^+$ and $F^-$, where ($F^+ = F^-$). To stabilize the pendulum, the torque applied at point $O$ generated by the force $F^+$ must be greater than the sum of the torques generated by the gravitational forces $P$ and $F^-$.

## 2.3 Real-life applications

In this section, we present some applications of the IWIP in real systems in use or systems currently under study. These systems rely on the same fundamental concept of IWIP stabilization, where the principles of balance and oscillatory motion are applied to maintain stability. The concept of stabilization is widely used in various real-world systems, from control mechanisms in engineering to emerging technologies currently under study. The study in [14] presents a mechanical system based on the IWIP used for a Self-Balancing Electric Motorcycle (SBEM). The system is illustrated in Fig. 2.2. The inertia wheel in this case is used to generate reactive torque for stabilization. The same concept is also studied in [15], namely a bicycle robot. The whole system must be capable of maintaining its balance while stationary and also following a pre-defined trajectory at a desired velocity without losing stability. The core mechanism involves an inertia wheel connected to an electric actuator, which can spin in either direction at a controlled speed to keep the motorcycle upright. When the motor

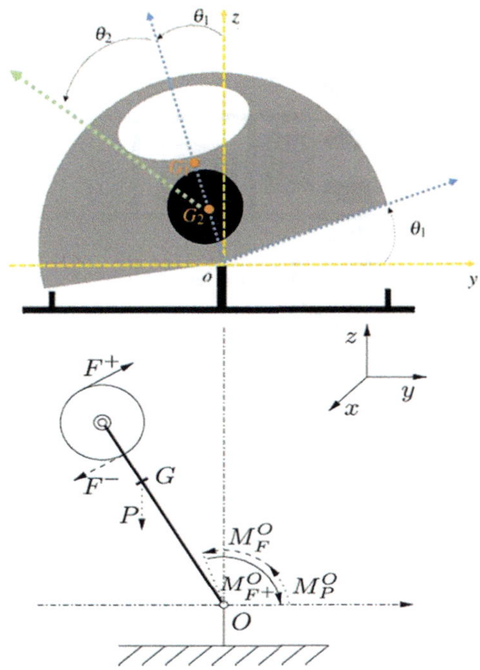


**FIGURE 2.1** Representation of the system: $\theta_1$ denotes the unactuated angular position of the pendulum, $\theta_2$ is the actuated angular position of the inertia wheel.

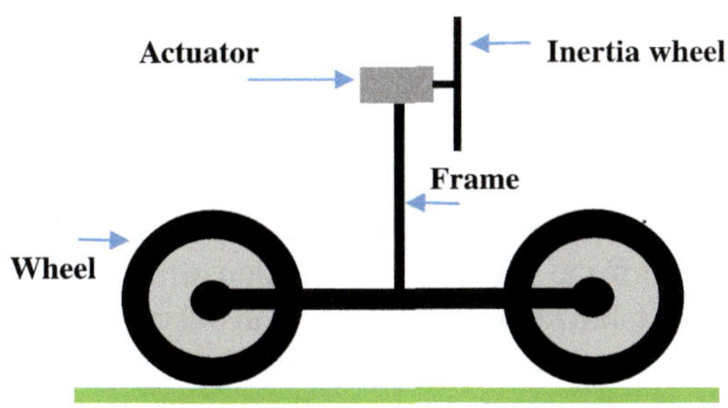


**FIGURE 2.2** Self-balancing electric motorcycle.

exerts torque on the inertia wheel, the wheel generates an equal and opposite torque on the motorcycle itself. By applying an appropriate control strategy, this interaction allows the bicycle to remain balanced. The stabilizing solution based on an inertia wheel can be proposed to avoid accidents when losing balance and control over the whole system. The control part of this system is extremely important. It is inherently an unstable system, and its nonlinear characteristics make it difficult to control. This complexity requires advanced techniques, such as for instance reinforcement learning as proposed in [16].

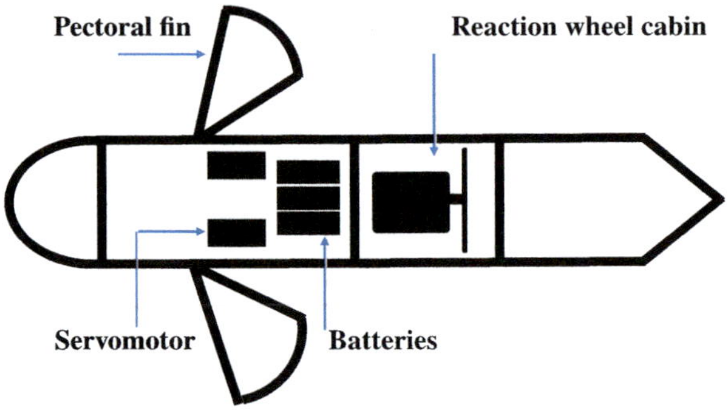


**FIGURE 2.3** Illustration of the use of a reaction wheel to stabilize the robotic fish.

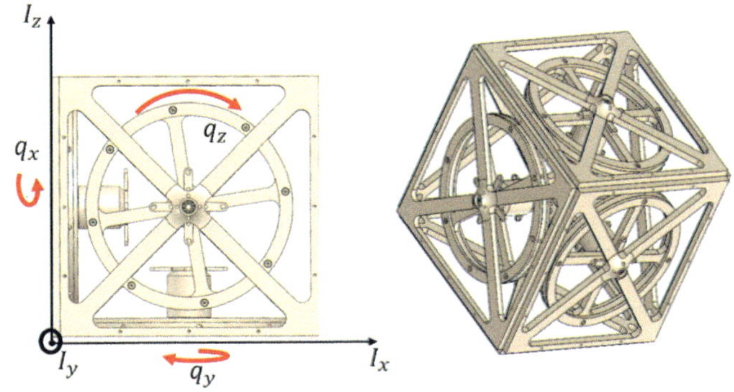


**FIGURE 2.4** 3D inertia wheel pendulum.

The authors in [18] proposed a solution to govern the roll stability of the robotic fish using a reaction wheel (see Fig. 2.3). The latter is employed for the attitude stabilization of the system. The reaction wheel can produce counterclockwise torque to stabilize the fish body and avoid disturbances.

Another innovative technology namely the Cubli system or self-balancing cube [19] (see Fig. 2.4), is composed of three reaction wheels mounted orthogonally to each other [21]. Each reaction wheel is a motor-driven wheel that can spin at high speeds. By controlling the speed and direction of the spinning wheels, the Cubli manipulates its orientation and movement. At high angular velocities, the Cubli is able to jump up from a resting position [20].

The authors in [22] proposed the design of an interesting one-wheel Cubli presented in Fig. 2.5. It is a 3D inverted pendulum achieving balancing using a single reaction wheel. It can serve as a challenging platform for research and education in the field of control.

The reaction wheel is also used for attitude control of spacecrafts [35–37]. Reaction wheels produce torque by spinning a wheel within the satellite [38]. An illustration is presented in Fig. 2.6, where the wheel is actuated using solar

FIGURE 2.5 The One-Wheel Cubli system.

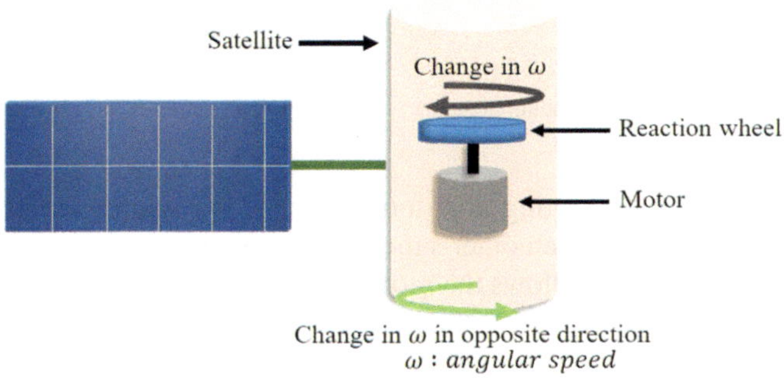


FIGURE 2.6 Illustration of a satellite with a reaction wheel.

panel electric energy. Due to the conservation of angular momentum, as the wheel rotates in one direction, the satellite turns in the opposite direction.

The inertia wheel and the gyroscope [17] are two mechanical systems that use similar physical principles, but they are designed for slightly different functions.

Both inertia wheel and gyroscope conserve rotational motion. Once they are set in motion, they tend to continue spinning unless an external force slows or stops them. This property is used for stabilization. Typically, an inertia wheel is a disk or wheel mounted on a fixed axis. It rotates about a single axis to store and release energy, regardless of movement on other axes. A gyroscope is often situated on a mount (called a gimbal) that allows it to rotate freely about multiple axes. This allows the gyroscope to maintain stability and counteract changes in orientation in different directions (see Fig. 2.7).

A gyroscopic stabilizer uses this principle to stabilize an object or vehicle against unwanted movement [24,25,27,28]. For example, in a boat, a gyroscopic stabilizer can reduce the rolling caused by waves. When the ship begins to rock, the gyroscope compensates for this movement by creating a precession force, which acts in the opposite direction, thus reducing the rocking. An interesting review on different gyrostabilizer vehicular technology can be found in [23]. Innovative electric vehicles, particularly self-balancing electric motorcycles as designed by [29], use two gyroscopes mounted at the base of the vehicle to

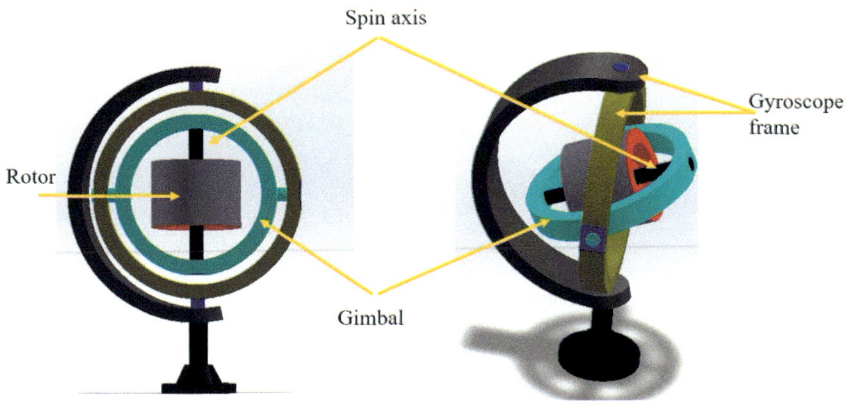


FIGURE 2.7 Gyroscope component.

maintain balance by generating high torque. These gyroscopes are arranged to counter-rotate, mirroring each other's motion. This configuration not only keeps the system stable but also allows the motorcycle to lean and turn left or right by adjusting the torque generated by the gyroscopes.

Gyroscopic actuators also significantly influence balance and gait kinematics, potentially improving postural control and stability. A study in [26] proposed a wearable gyroscopic actuator, namely the GyroPack, tested on non-impaired individuals. The study assessed the impact of wearing the weight wearable gyroscopic on balance and gait. An illustration of this system is presented in Fig. 2.8. Postural control can be influenced by manipulating the angular momentum. To adjust the latter, control moment gyroscopes (CMGs) are commonly used in wearable devices [30]. Other designed prototypes can be found in the literature, such as [31–33]. However, the reduction in device mass and the need for versatile control approaches emphasize their necessity for practical and clinical use.

## 2.4 Mathematical modeling of the system

The process of mathematical modeling for mechanical systems involves using a mathematical framework to represent and describe the behavior and dynamics of these systems. In this section, we will present step by step how to create the model of our system using the Euler–Lagrange method. Furthermore, we will represent the system in open loop, the linearized system, as well as the Hamiltonian representation.

### 2.4.1 Dynamic model

The IWIP is classified as a system with a flatness property [34] and kinetic symmetry with a dynamics including a constant inertia matrix.

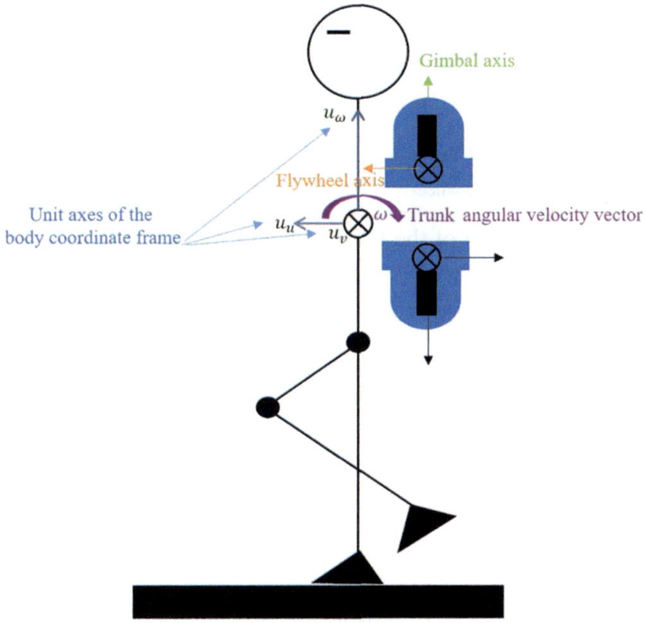


**FIGURE 2.8** Weight wearable gyroscopic actuators for human balance equipped with two CMGs.

The dynamic model is obtained through the application of the Euler–Lagrange formalism. This yields the equations describing the evolution of the dynamics of the system coordinates $\theta_1$ and $\theta_2$ over time, as a function of the torque generated by the actuator acting on the inertia wheel. The kinetic energy (denoted $T$) and potential energy (denoted $V$) of the system are expressed in terms of the generalized coordinates. The Lagrangian of the system is denoted as $L = T - V$. The Euler–Lagrange equations are then written as follows:

$$\frac{d}{dt}\left(\frac{\partial L}{\partial \dot{\theta}_j}\right) - \frac{\partial L}{\partial \theta_j} = \tau_j \quad \text{for} \quad j = 1, 2 \tag{2.1}$$

where $\mathbf{q} = \begin{pmatrix} q_1 \\ q_2 \end{pmatrix} = \begin{pmatrix} \theta_1 \\ \theta_2 \end{pmatrix}$ is the vector of generalized coordinates, $\tau_\mathbf{i} = \begin{pmatrix} C_1 \\ C_2 \end{pmatrix}$ is the vector of generalized forces represented by torques in our case. The total kinetic energy of the system is decomposed into the energy of the pendulum and the inertia wheel:

$$T = T_{\text{pendulum}} + T_{\text{inertia-wheel}}$$
$$T_{\text{pendulum}} = \frac{1}{2}\left(m_1 l_1^2 + i_1\right)\dot{\theta}_1^2$$
$$T_{\text{inertia-wheel}} = \frac{1}{2} m_2 l_2^2 \dot{\theta}_2^2 + \frac{1}{2} i_2 \left(\dot{\theta}_1 + \dot{\theta}_2\right)^2$$

The total potential energy of the system is also decomposed into the energy of the pendulum and the inertia wheel:

$$V = V_{\text{pendulum}} + V_{\text{inertia-wheel}} \tag{2.2}$$

$$V_{\text{pendulum}} = m_1 l_1 g \cos\theta_1 \tag{2.3}$$

$$V_{\text{inertia-wheel}} = m_2 l_2 g \cos\theta_1 \tag{2.4}$$

The Lagrangian expression of the IWIP is then given by:

$$L = \frac{1}{2}(I_1\dot{\theta}_1^2 + i_2(\dot{\theta}_1 + \dot{\theta}_2)^2) - m_0 g \cos(\theta_1) \tag{2.5}$$

We now apply the Euler–Lagrange equation. For $j = 1$, we derive the expression for $\frac{\partial L}{\partial \dot{\theta}_1}$:

$$\frac{\partial L}{\partial \dot{\theta}_1} = I_1\dot{\theta}_1 + I_2(\dot{\theta}_1 + \dot{\theta}_2) \tag{2.6}$$

Now, we compute $\frac{\partial L}{\partial \theta_1}$:

$$\frac{\partial L}{\partial \theta_1} = \frac{\partial}{\partial \theta_1}(-m_0 g \cos(\theta_1)) = m_0 g \sin(\theta_1) \tag{2.7}$$

By applying the Euler–Lagrange equation for $j = 1$, we obtain:

$$\frac{d}{dt}\left(I_1\dot{\theta}_1 + i_2(\dot{\theta}_1 + \dot{\theta}_2)\right) - m_0 g \sin(\theta_1) = C_1 \tag{2.8}$$

Taking the time derivative:

$$I_1\ddot{\theta}_1 + i_2(\ddot{\theta}_1 + \ddot{\theta}_2) - m_0 g \sin(\theta_1) = C_1 \tag{2.9}$$

Simplifying Eq. (2.9), we obtain:

$$(I_1 + i_2)\ddot{\theta}_1 + i_2\ddot{\theta}_2 - m_0 g \sin(\theta_1) = C_1 \tag{2.10}$$

Following the same steps for $j = 2$, we compute $\frac{\partial L}{\partial \dot{\theta}_2}$:

$$\frac{\partial L}{\partial \dot{\theta}_2} = \frac{\partial}{\partial \dot{\theta}_2}\left(\frac{1}{2}i_2(\dot{\theta}_1 + \dot{\theta}_2)^2\right) = i_2(\dot{\theta}_1 + \dot{\theta}_2) \tag{2.11}$$

Since $L$ does not depend explicitly on $\theta_2$, we have:

$$\frac{\partial L}{\partial \theta_2} = 0 \tag{2.12}$$

We now apply the Euler–Lagrange equation for $j = 2$, to obtain:

$$\frac{d}{dt}\left(i_2(\dot{\theta}_1 + \dot{\theta}_2)\right) = C_2 \tag{2.13}$$

**TABLE 2.1 Dynamical parameters of the system.**

| Notation | Description | Units |
|---|---|---|
| $i_1$ | Pendulum Inertia | [kg m$^2$] |
| $i_2$ | Wheel Inertia | [kg m$^2$] |
| $l_1$ | Distance $OG_1$ | [m] |
| $l_2$ | Distance $OG_2$ | [m] |
| $m_1$ | Body mass | [kg] |
| $m_2$ | Mass of the wheel | [kg] |

Taking the time derivative, we obtain:

$$i_2(\ddot{\theta}_1 + \ddot{\theta}_2) = C_2 \tag{2.14}$$

The resulting dynamic model of an IWIP obtained using Lagrange's method is expressed as follows:

$$(I_1 + i_2)\ddot{\theta}_1 + i_2\ddot{\theta}_2 - m_0 g \sin(\theta_1) = C_1 = 0 \tag{2.15}$$

$$i_2(\ddot{\theta}_1 + \ddot{\theta}_2) = C_2 = \tau \tag{2.16}$$

where $\theta = [\theta_1, \dot{\theta}_1, \dot{\theta}_2]^T$ is the state vector including, respectively, the angular position of the pendulum, the velocity of the pendulum, and the velocity of the inertia wheel, $[\ddot{\theta}_1,\ \ddot{\theta}_2]$ are, respectively, the acceleration of the pendulum and rotating wheel, $g = 9.8$ m/s$^2$ is the constant of gravitational acceleration, $\tau$ denotes the applied torque exerted by the actuator and $I_1 = m_1(l_1)^2 + m_2(l_2)^2 + i_1$, $m_0 = m_1 l_1 + m_2 l_2$. The torque $C_2$ is the result of the torque induced by the actuator and the torque induced by possible disturbances, while $C_1$ is a torque purely resulting from disturbances. It is important to note that through the torque $C_2$ the control acts on the system. The system dynamics is defined in terms of certain dynamical parameters summarized in Table 2.1.

We have considered the following remarks with the aim of model simplification. The DC motor model mounted on the wheel was not considered in this study. According to the physical device, the frictions on both actuated and unactuated joints are assumed to be low. Therefore the system's frictional forces were not considered in the dynamic modeling. The angular position of the inertia wheel $\theta_2$ was also not considered as a state variable, as it does not play a significant role in the Lagrange equations and the dynamics.

### 2.4.2 Open-loop system

In open-loop systems, external disturbances and control are reduced to zero (it is assumed that the system is not subject to disturbances). The equations of the

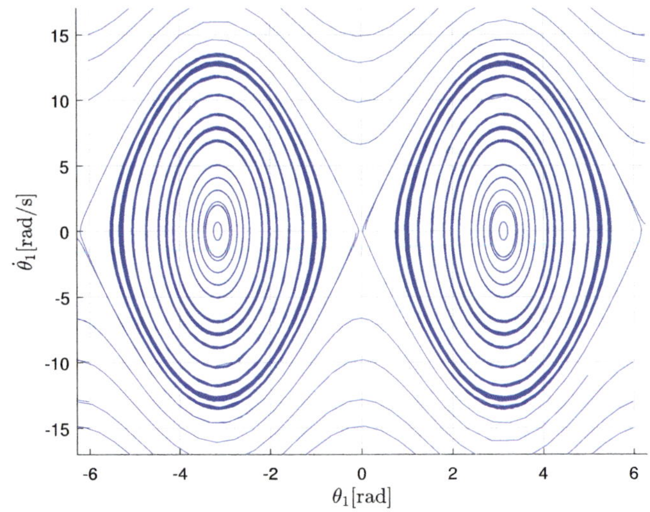


**FIGURE 2.9** Phase portrait of the open-loop system.

pendulum dynamics are then written as follows, after substituting the values of $C_1$ and $C_2$:

$$\ddot{\theta}_1 = \frac{m_0 g}{I_1} \sin\theta_1 \tag{2.17}$$

$$\ddot{\theta}_2 = -\ddot{\theta}_1 \tag{2.18}$$

We observe from these equations that the position $\theta_2$ and velocity $\dot{\theta}_2$ of the inertia wheel have no influence on the evolution of the system. It is an autonomous system whose evolution depends only on its parameters and initial condition. The evolution of the inertia wheel trajectory is completely determined by that of the pendulum (it is the exact opposite, except for a translation). The open-loop system has two equilibrium points: the upright position $(\theta_1, \theta_2) = (0, 0)$, which is unstable, and the position $(\theta_1, \theta_2) = (\pi, 0)$, which is stable. The phase diagram of the resulting autonomous open-loop system is depicted in Fig. 2.9. The trajectories are plotted for different initial conditions. We can observe the two equilibrium points as well as the different orbits of the system. Below a certain initial velocity condition of the pendulum, the trajectories are presented by some oscillations around the stable equilibrium point $(\theta_1, \theta_2) = (\pi, 0)$. Above this initial velocity, the pendulum rotates indefinitely around the passive joint in the same direction as the initial velocity.

### 2.4.3 Port-Hamiltonian model

Port-Hamiltonian systems theory provides a systematic methodology for the modeling, simulation, and control of systems including robots and mechanical

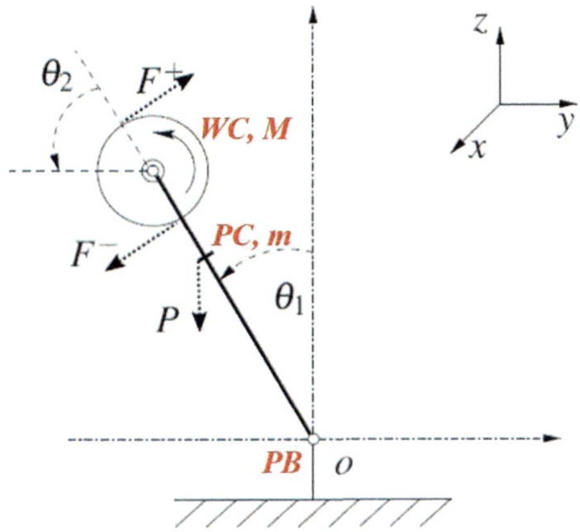


**FIGURE 2.10** Representation of the inertia wheel inverted pendulum.

systems [39]. Port-Hamiltonian models have been used to design controllers for flexible beams in medical endoscopes [40], a cable driven robot [41], unmanned surface vessels [42], and underactuated systems [8,11,43–46]. This model representation is commonly employed to stabilize underactuated systems by shaping the system's energy and maintaining passivity. The latter means that the system does not generate energy by itself, which guarantees stability under certain conditions. The equations of motion of an underactuated mechanical system can be written in Port-Controlled Hamiltonian form (PCH) as:

$$\begin{bmatrix} \dot{q} \\ \dot{p} \end{bmatrix} = \begin{bmatrix} 0_{n\times n} & I_n \\ I_n & 0_{n\times n} \end{bmatrix} \nabla H(q,p) + \begin{bmatrix} 0_{n\times m} \\ G(q) \end{bmatrix} u \tag{2.19}$$

$$y = G(q)^T \nabla_p H \tag{2.20}$$

where $(q, p) \in \mathbb{R}^n$ are, respectively, the generalized position and momenta, $u \in \mathbb{R}^m$ is the control input vector, $G(q) : \mathbb{R}^n \to \mathbb{R}^{n\times m}$ is the input matrix with $rank(G) = m < n$, $y$ is the output vector, and $H : \mathbb{R}^n \times \mathbb{R}^n \to \mathbb{R}$ is the total energy. This last one is defined as the sum of the kinetic and potential energy, expressed as follows:

$$H(q,p) := \frac{1}{2} p^T M^{-1}(q) p + V(q) \tag{2.21}$$

where $M : \mathbb{R}^n \to \mathbb{R}^{n\times n}$ is a positive-definite inertia matrix, $M(q) = M(q)^T > 0$ and $V(q) : \mathbb{R}^n \to \mathbb{R}$ is the potential energy.

The dynamic model of IWIP (see Fig. 2.10) that is obtained using the Euler–Lagrange method as described in Eqs. (2.15) and (2.16) can be expressed in the following form:

$$\begin{bmatrix} a + I_{WC} & I_{WC} \\ I_{WC} & I_{WC} \end{bmatrix} \begin{bmatrix} \ddot{\theta}_1 \\ \ddot{\theta}_2 \end{bmatrix} - \begin{bmatrix} bg\sin(\theta_1) \\ 0 \end{bmatrix} = \begin{bmatrix} 0 \\ u \end{bmatrix} \tag{2.22}$$

where $\theta = [\theta_1, \theta_2]^T$, is the vector of generalized positions, $u$ is the torque generated by the actuator and acting on the inertia wheel. In the above model, the terms related to the friction in the pendulum joint are also neglected. $a = ML^2 + I_{PB}$ and $b = ml + ML$. The dynamic parameters of the system $I_{PC}$ and $I_{PB}$ denote the rotational inertia of the pendulum about the pendulum center of mass (PC) and the rotational inertia of pendulum about its base (PB), respectively. $I_{WC}$ represents the rotational inertia of the wheel about its center of mass (WC). $l$ denotes the length from the pendulum base (PB) to the pendulum center of mass (PC), the length from the pendulum base (PB) to the wheel center of mass (WC) is denoted by $L$, $m$, and $M$ are the mass of the pendulum and wheel, respectively, and $g$ is the constant of gravitational acceleration.

The first step is to transform the system model in PCH form. Consider the following change of coordinates

$$q_1 = \theta_1 \tag{2.23}$$

$$q_2 = \theta_1 + \theta_2 \tag{2.24}$$

$$p_1 = a\dot{q}_1 \tag{2.25}$$

$$p_2 = I_{WC}\dot{q}_2 \tag{2.26}$$

We obtain the following simplified model with a diagonal inertia matrix:

$$\begin{pmatrix} a & 0 \\ 0 & I_{WC} \end{pmatrix} \begin{pmatrix} \ddot{q}_1 \\ \ddot{q}_2 \end{pmatrix} - \begin{pmatrix} bg\sin(q_1) \\ 0 \end{pmatrix} = \begin{pmatrix} -1 \\ 1 \end{pmatrix} u \tag{2.27}$$

The obtained model is rewritten through Hamilton's equations of motion as

$$\begin{pmatrix} \dot{q}_1 \\ \dot{q}_2 \\ \dot{p}_1 \\ \dot{p}_2 \end{pmatrix} = \begin{pmatrix} \frac{p_1}{a} \\ \frac{p_2}{I_{WC}} \\ bg\sin(q_1) - u \\ u \end{pmatrix} \tag{2.28}$$

where $q = [q_1, q_2]^T$ and $p = [p_1, p_2]^T$ are the generalized positions and momenta, respectively.

### 2.4.4 Linearized model

The state representation of the linearized system is obtained by linearizing its dynamic model around the unstable equilibrium point, where the pendulum is in the upright vertical position. The state equations are expressed in the form:

$$\dot{X} = AX + BU$$
$$Y = CX + DU$$

with $X^T = \begin{bmatrix} \theta_1 & \dot{\theta}_1 & \dot{\theta}_2 \end{bmatrix}$ and $U = \begin{bmatrix} C_1 \\ C_2 \end{bmatrix}$

$$\begin{bmatrix} \dot{\theta}_1 \\ \ddot{\theta}_1 \\ \ddot{\theta}_2 \end{bmatrix} = \begin{bmatrix} 0 & 1 & 0 \\ A_{21} & 0 & 0 \\ A_{31} & 0 & 0 \end{bmatrix} \begin{bmatrix} \theta_1 \\ \dot{\theta}_1 \\ \dot{\theta}_2 \end{bmatrix} + \begin{bmatrix} 0 \\ B_1 \\ B_2 \end{bmatrix} C_2$$

$$Y = \begin{bmatrix} C_{11} & 0 & 0 \end{bmatrix} \begin{bmatrix} \theta_1 \\ \dot{\theta}_1 \\ \dot{\theta}_2 \end{bmatrix}$$

where $A_{21} = \frac{m_0 g}{I_1}$, $A_{31} = -\frac{m_0 g}{I_1}$, $B_1 = \frac{-1}{I_1}$, $B_2 = \frac{I_1 + i_2}{I_1 i_2}$, $C_{11} = 1$.

The poles of the open-loop system are calculated by finding the solutions to the characteristic equation $\det(A - \lambda I) = 0$. In the case of the IWIP, the characteristic equation is: $\lambda(\lambda^2 - \frac{m_0 g}{I}) = 0$ and the poles are $\lambda_1 = 0$, $\lambda_2 = \frac{m_0 g}{I_1}$, and $\lambda_3 = -\frac{m_0 g}{I_1}$. The presence of poles with zero or positive real parts confirms that the system is unstable. A system is said to be completely controllable if and only if the rank of its controllability matrix $C_0 = [B\ AB\ \ldots\ A^{n-1}B]$ is equal to the dimension of the system ($n$). In this case, $C_0 = [B\ AB\ A^2B]$ and $n = 3$. The rank of $C_0$ is $n = 3$, so the system is completely controllable. A system is completely observable if and only if the rank of its observability matrix $O = \begin{bmatrix} C^T & [CA]^T & \ldots & [CA^{n-1}]^T \end{bmatrix}^T$ is equal to the dimension of the system. In our case, the rank of $O$ is $n = 3$, so the system is completely observable.

## 2.5 Experimental setup and implementation issues

Experiments can reveal complex behaviors in nonlinear underactuated systems that may not be fully understood through numerical simulation. An immediate observation of system behavior under various conditions, allows a quick adjustment of the control gain and improves the accuracy of controllers. For these reasons, the description and understanding of the testbed are crucial for conducting effective experiments and to ensure that the experimental setup is accurately aligned with theoretical models and objectives. In this section, the experimental setup of the IWIP and some implementation issues are discussed. This will include a detailed overview of the physical testbed used for the experiments, the mechanical and electrical parts of the system, and the methodologies applied to test the robustness of any implemented control approach.

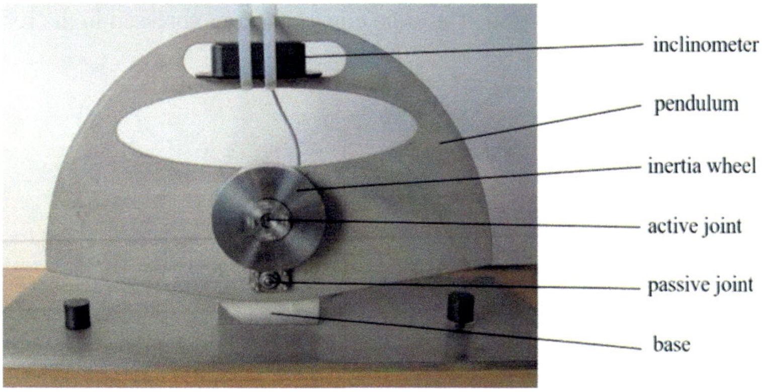


**FIGURE 2.11** View of the real inverted pendulum stabilized by an inertia wheel.

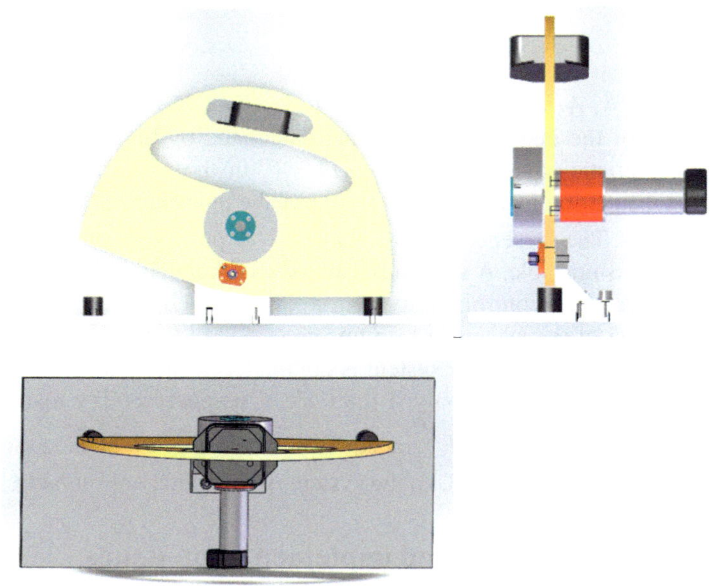

**FIGURE 2.12** Front view, top view, and side view of the IWIP.

## 2.5.1 Mechanical part

A mechanical view of the IWIP is presented in Figs. 2.11 and 2.12. Due to mechanical stops, the pendulum is constrained to remain within the range $[-10°, 10°]$.

The system is composed of a pendulum connected to a fixed support via an unactuated pivot link, controlled by an actuated inertia wheel. The descriptions of the motor and sensors are presented in the next subsection.

**TABLE 2.2** Actuator's specifications.

| Notation | Value |
|---|---|
| Nominal voltage | 24 V |
| Nominal torque | 95.6 mN m |
| Nominal current | 7.61 A |

### 2.5.2 Electrical part

**Actuator and encoder:** The size and selection of the actuator motor was closely related to the size of the inertia wheel. The selected motor is the Maxon EC-powermax 30. It is a motor coupled with a gearbox and an incremental encoder.

The role of a gearbox is to modify the output characteristics of a motor, typically by altering the speed and torque [47]. It achieves this by using gears to either increase the torque while reducing the speed or increase the speed while reducing the torque, depending on the gear ratio.

The incremental encoder, on the other hand, allows the measurement of its angular position and, consequently, that of the inertia wheel. Some of the motor's specifications are described in Table 2.2. For more details on the dimensions, the reader can refer to the corresponding motor datasheet.

**Inclinometer:** An inclinometer (FAS-G of Micro strain) is mounted on the body of the pendulum. It provides an analog voltage output proportional to the inclination of the pendulum with respect to the vertical axis. This was the only solution to measure the angular position, as the pendulum is not actuated, and an encoder cannot be used in this case.

**Power Supply:** The entire system is powered by a voltage of 12 V.

**Computer:** The PC used for control runs a real-time operating system called RTX (Real-time Extension for Windows) developed by Ardence. Visual C++ is utilized for programming the control algorithms, while MATLAB/SIMULINK® is employed for result visualization. The sampling time is configured to 6 ms, ensuring accurate and timely data collection.

**Interface Card:** The communication between the computer and the physical system is ensured by an interface card. It enables the data exchange between the computer and sensors of the testbed.

**Variable Frequency Drive (VFD)** is an electronic device used to control the speed and torque of an electric motor by varying the frequency and voltage of the electrical power supplied to the motor.

### 2.5.3 Software description

RTX Ardence (IntervalZero RTX/RTX64) transforms Windows into a real-time operating system (RTOS) by adding a Real-Time Sub-system (RTSS). This subsystem enables real-time tasks to execute deterministically, independent of the

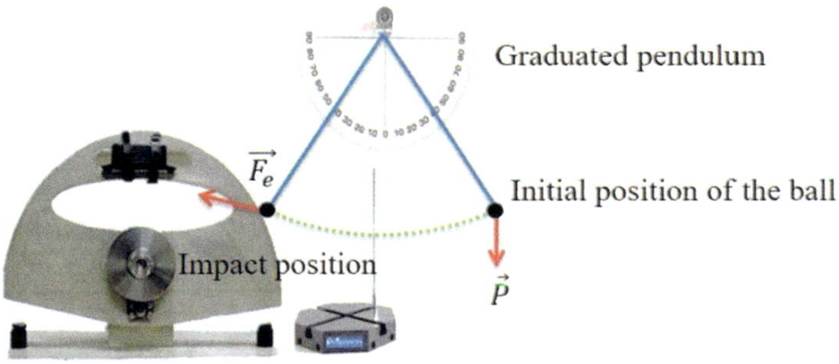


**FIGURE 2.13** Illustration of punctual disturbance.

Windows scheduler. Visual Studio is the integrated development environment (IDE) used to develop, compile, and debug real-time applications that run on the RTSS provided by RTX Ardence. Once the project is compiled in Visual Studio C++, it generates an executable .RTSS file specifically designed for the real-time sub-system (RTSS) provided by RTX. Running this .RTSS file initiates the real-time experimentation process. Throughout the execution, data from real-time variables, sensor readings, and system performance are recorded. Once the execution is complete, these data points are saved in .MAT files, which can then be conveniently imported into MATLAB. It enables clear visualization and analysis of the experimental results, helping to evaluate the system's behavior and performance in real time.

### 2.5.4 Description of the evaluation scenarios

When an experiment is conducted in the nominal case, this refers to a scenario in which the system operates under ideal conditions. However, disturbances in real-time control systems are a significant challenge in control theory, impacting system performance and stability. In this section, we will describe some scenarios used to demonstrate the robustness and effectiveness of the implemented controllers, particularly in the presence of disturbances. The first case we will discuss is the punctual disturbance and the second case we will discuss is persistent disturbances.

#### 2.5.4.1 *Case of rejection of punctual disturbances*

In this case, the pendulum is perturbed with a classic graduated pendulum suspended from a fixed support with a mass ball. To perturb the system and test several controllers under the same conditions, the disturbing mass of the pendular system was released from the same initial position (e.g., 20°) and at the same time (e.g., 5 s). This approach ensures that the same external force $F_e$ is applied at the point of impact on the pendulum. An illustration of the scenario is depicted in Fig. 2.13.

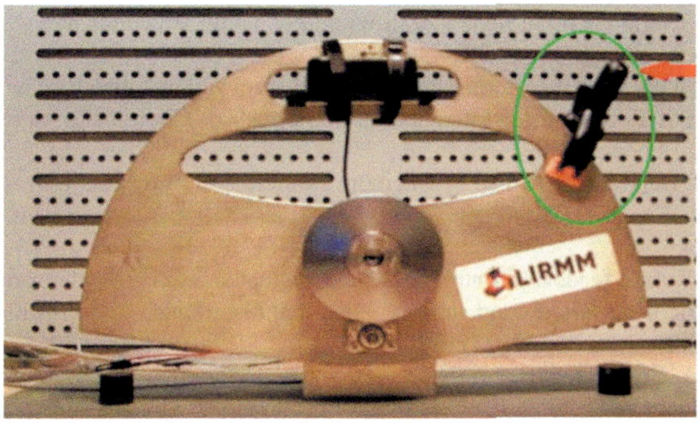


**FIGURE 2.14** Illustration of persistent disturbance.

#### 2.5.4.2 Case of persistent disturbances

In this scenario, a constant force, representing a persistent disturbance, is applied to the inverted pendulum through the addition of a mass attached to its body. This additional mass creates a continuous external influence, challenging the stability and control of the pendulum. An illustration of the scenario is depicted in Fig. 2.14. By examining the response of the system to this perturbation, we can evaluate the effectiveness of the implemented controllers in maintaining balance and achieving the desired performance under these conditions.

## 2.6 Conclusion

The inertia wheel inverted pendulum inherits the main control challenges of more complex underactuated mechanical systems, especially those actuated by one or more inertia-wheel gyrostabilizers. It is a complex and sophisticated system that requires precise control to operate effectively. It has proven to be a highly effective technology for stabilizing vehicular systems and it is widely used in a variety of applications. The system's description and all the mathematical modeling and representation of the system are detailed in this chapter and will severe for the case study of subsequent chapters. Real-time experiments in control design provide immediate feedback, allowing for dynamic analysis and rapid adjustments to control gains, which enhances system performance and robustness.

## References

[1] O. Boubaker, R. Iriarte, The inverted pendulum in control theory and robotics: from theory to new innovations, The Inverted Pendulum in Control Theory and Robotics (2017) 1–395.

[2] A. Lukowska, Simple human walking modelling, in: Proceedings of the International Conference BIOMDLORE, 2018.

[3] S. Kajita, F. Kanehiro, K. Kaneko, K. Fujiwara, K. Yokoi, H. Hirukawa, Biped walking pattern generation by a simple three-dimensional inverted pendulum model, Advanced Robotics 17 (2003) 131–147.

[4] Z. Hazem, Z. Bingul, Comprehensive review of different pendulum structures in engineering applications, IEEE Access 11 (2023) 42862–42880.
[5] B. Garabédian, M. Benoit, S. Krut, A futuristic monorail tramway stabilized by an inertia wheel, in: Proceedings of the 2007 IEEE International Conference on Control and Automation, 2007, pp. 1581–1586.
[6] N. Khraief Haddad, S. Belghith, H. Gritli, A. Chemori, From Hopf bifurcation to limit cycles control in underactuated mechanical systems, International Journal of Bifurcation and Chaos 27 (7) (2017) 1750104.
[7] H. Gritli, S. Belghith, Robust feedback control of the underactuated inertia wheel inverted pendulum under parametric uncertainties and subject to external disturbances: LMI formulation, Journal of the Franklin Institute 355 (2017) 9150–9191.
[8] H. Gritli, N. Khraief, A. Chemori, S. Belghith, Self-generated limit cycle tracking of the underactuated inertia wheel inverted pendulum under IDA-PBC, Nonlinear Dynamics 89 (3) (2017) 2195–2226, https://doi.org/10.1007/s11071-017-3578-y.
[9] A. Hfaiedh, A. Chemori, A. Abdelkrim, Rise controller for class I of underactuated mechanical systems: design and real-time experiments, in: Proceedings of the 3rd International Conference on Electromechanical Engineering (ICEE), Skikda, Algeria, 2018.
[10] A. Hfaiedh, A. Chemori, A. Abdelkrim, Disturbance observer-based super-twisting control for the inertia wheel inverted pendulum, in: Proceedings of the 17th International Multi-Conference on Systems, Signals and Devices (SSD), Sfax, Tunisia, 2020, pp. 751–756.
[11] A. Hfaiedh, A. Chemori, A. Abdelkrim, Stabilization of the inertia wheel inverted pendulum by advanced IDA-PBC based controllers: comparative study and real-time experiments, in: Proceedings of the 17th International Multi-Conference on Systems, Signals and Devices (SSD), Sfax, Tunisia, 2020, pp. 757–764.
[12] A. Hfaiedh, A. Chemori, A. Abdelkrim, Observer-based robust integral of the sign of the error control of class I of underactuated mechanical systems: theory and real-time experiments, Transactions of the Institute of Measurement and Control (2021).
[13] A. Hfaiedh, A. Abdelkrim, Revisited adaptive sliding mode control of underactuated mechanical systems with real-time experiments, International Journal of Modelling, Identification and Control 37 (3/4) (2021) 344–353.
[14] Q.K. Ho, C.B. Pham, Study on inertia wheel pendulum applied to self-balancing electric motorcycle, in: Proceedings of the 2018 4th International Conference on Green Technology and Sustainable Development (GTSD), Ho Chi Minh City, Vietnam, 2018, pp. 687–692.
[15] K. Kanjanawanishkul, LQR and MPC controller design and comparison for a stationary self-balancing bicycle robot, Kybernetika 51 (1) (2015) 173–191.
[16] S. Vadlamudi, K.V. Lakshmi, A. Yaramala, C. Uppalapati, P.K. Kolluri, Self balancing motorcycle using reinforcement learning, in: Proceedings of the 2024 International Conference on Emerging Systems and Intelligent Computing (ESIC), Bhubaneswar, India, 2024, pp. 691–696.
[17] W. Burger, A.G. Corbet, The gyroscope, in: Ship Stabilizers, Pergamon, 1966, pp. 17–30.
[18] P. Zhang, Z. Wu, H. Dong, M. Tan, J. Yu, Reaction-wheel-based roll stabilization for a robotic fish using neural network sliding mode control, IEEE/ASME Transactions on Mechatronics 25 (4) (2020) 1904–1911.
[19] J. Mayr, F. Spanlang, H. Gattringer, Mechatronic design of a self-balancing three-dimensional inertia wheel pendulum, Mechatronics 30 (2015) 1–10.
[20] M. Gajamohan, M. Merz, I. Thommen, R.D'Andrea, The Cubli: a cube that can jump up and balance, in: 2012 IEEE/RSJ International Conference on Intelligent Robots and Systems, 2012, pp. 3722–3727.
[21] M. Muehlebach, G. Mohanarajah, R.D'Andrea, Nonlinear analysis and control of a reaction wheel-based 3D inverted pendulum, in: 52nd IEEE Conference on Decision and Control, 2013, pp. 1283–1288.
[22] M. Hofer, M. Muehlebach, R.D'Andrea, The One-Wheel Cubli: a 3D inverted pendulum that can balance with a single reaction wheel, Mechatronics 91 (2023) 102965, https://doi.org/10.1016/j.mechatronics.2023.102965.

[23] N.C. Townsend, R.A. Shenoi, Gyrostabilizer vehicular technology, ASME Applied Mechanics Reviews 64 (1) (2011) 010801, https://doi.org/10.1115/1.4004837.
[24] S-A. Kim, A study on the control for increasing output power section of gyro generation system, Transactions of the Korean Institute of Electrical Engineers 69 (4) (2020) 576–580, https://doi.org/10.5370/KIEE.2020.69.4.576.
[25] S. Spry, A. Girard, Gyroscopic stabilisation of unstable vehicles: configurations, dynamics, and control, Vehicle System Dynamics 46 (2008) 247–260, https://doi.org/10.1080/00423110801935863.
[26] B.T. Sterke, K.L. Poggensee, G.M. Ribbers, D. Lemus, H. Vallery, Light-weight wearable gyroscopic actuators can modulate balance performance and gait characteristics: a proof-of-concept study, Healthcare (Basel) 11 (21) (2023) 2841.
[27] N.C. Townsend, A.J. Murphy, R.A. Shenoi, A new active gyrostabiliser system for ride control of marine vehicles, Ocean Engineering 34 (11) (2007) 1607–1617, https://doi.org/10.1016/j.oceaneng.2006.11.004.
[28] A. Mahvan, A. Akbarzadeh, Gyrostabilized two wheeled inverted pendulum robot, in: 2015 3rd RSI International Conference on Robotics and Mechatronics (ICROM), Tehran, Iran, 2015, pp. 669–674.
[29] https://www.litmotors.com/.
[30] P. Romtrairat, C. Virulsri, P. Wattanasiri, P. Tangpornprasert, A performance study of a wearable balance assistance device consisting of scissored-pair control moment gyroscopes and a two-axis inclination sensor, Journal of Biomechanics 109 (2020) 109957.
[31] D. Lemus, A. Berry, S. Jabeen, et al., Controller synthesis and clinical exploration of wearable gyroscopic actuators to support human balance, Scientific Reports 10 (1) (2020) 10412.
[32] J. Chiu, A. Goswami, Design of a wearable scissored-pair control moment gyroscope (SP-CMG) for human balance assist, in: Proceedings of the ASME 2014 International Design Engineering Technical Conferences and Computers and Information in Engineering Conference. Volume 5A: 38th Mechanisms and Robotics Conference, Buffalo, New York, USA, 2014.
[33] H. Oya, Y. Fujimoto, Preliminary experiments for postural control using wearable-CMG, in: IECON 2017 - 43rd Annual Conference of the IEEE Industrial Electronics Society, Beijing, China, 2017, pp. 7602–7607.
[34] R. Olfati-Saber, Global stabilization of a flat underactuated system: the inertia wheel pendulum, in: Proceedings of the 40th IEEE Conference on Decision and Control (Cat. No. 01CH37228), vol. 4, 2001, pp. 3764–3765.
[35] B. Çolak, A. Özdemir, Ş Ötenkaya, B. Selamlar, M. Öncüler, S. Çetinkaya, First reaction wheel qualified in Türkiye, in: 2023 10th International Conference on Recent Advances in Air and Space Technologies (RAST), 2023, pp. 1–6.
[36] K. Ratheesh, S. Sreejith, T. Haridas, Field-oriented control for performance improvement in reaction wheels and implementation of algorithm in FPGA, in: Advances in Small Satellite Technologies, 2023, pp. 143–151.
[37] M. Mehrjardi, H. Sanusi, M. Ali, Developing a proposed satellite reaction wheel model with current mode control, in: 2015 International Conference on Space Science and Communication (IconSpace), 2015, pp. 410–413.
[38] D. Cheon, E. Jang, H. Oh, Satellite actuator balancing based on the disturbance measurement table data, Renewable Energy and Power Quality Journal (2012) 601–604.
[39] A. Schaft, D. Jeltsema, Port-Hamiltonian Systems Theory: an Introductory Overview. Now Foundations and Trends, 2014.
[40] A. Mattioni, Y. Wu, H. Ramirez, Y. Gorrec, A. Macchelli, Modelling and control of a class of lumped beam with distributed control, IFAC-PapersOnLine 51 (2018) 217–222, 6th IFAC Workshop on Lagrangian and Hamiltonian Methods for Nonlinear Control (LHMNC 2018).
[41] M. Harandi, S. Khalilpour, H. Taghirad, Robust IDA-PBC for a spatial underactuated cable driven robot with bounded inputs, in: 2021 29th Iranian Conference on Electrical Engineering, ICEE 2021, 2021, pp. 689–694.

[42] W. Zhou, Z. Xu, Y. Wu, J. Xiang, Y. Li, Energy-based trajectory tracking control of underactuated unmanned surface vessels, Ocean Engineering (2023) 288.
[43] R. Ortega, M. Spong, F. Gómez-Estern, G. Blankenstein, Stabilization of a class of underactuated mechanical systems via interconnection and damping assignment, IEEE Transactions on Automatic Control 47 (2002) 1218–1233.
[44] N. Haddad, A. Chemori, J. Pena, S. Belghith, Stabilization of inertia wheel inverted pendulum by model reference adaptive IDA-PBC: from simulation to real-time experiments, in: 2015 3rd International Conference on Control, Engineering & Information Technology (CEIT), 2015, pp. 1–6.
[45] N. Khraief-Haddad, A. Chemori, S. Belghith, Robustness enhancement of IDA-PBC controller in stabilising the inertia wheel inverted pendulum: theory and real-time experiments, International Journal of Control 91 (2018) 2657–2672.
[46] M. Ryalat, D. Laila, A robust IDA-PBC approach for handling uncertainties in underactuated mechanical systems, IEEE Transactions on Automatic Control 63 (2018) 3495–3502.
[47] S. Suresh, M. Srinivas, V. Naidu, Vibration analysis of gearbox fault diagnosis using DWT and statistical features, Journal of Engineering Research (Kuwait) 10 (2022) 156–165.

Part II

# Control solutions for the stabilization problem

Chapter 3

# A revisited adaptive sliding mode control scheme

Contents



## 3.1 Introduction

Due to various reasons, some discrepancies may arise between the physical system and its derived mathematical model. Such differences exist due to simplifications considered on the developed model, the system parameters that may vary over time by degradation or environmental condition, uncertainties in measurement, noisy sensors, etc. Robust controls are typically developed in the literature to mitigate the effect of such unmodeled dynamics. However, this mismatch is more challenging to handle in systems with fewer control inputs than degrees of freedom. On the grounds of the limited control action over all degrees of freedom, the control problem becomes more challenging. This is due to the increased sensitivity to disturbances and parameter variation while considering their complex dynamics and stability analysis. To address such aspects, robust and advanced control methodologies are required. To deal with such control issues, a well-known robust approach in control design is the sliding mode control (SMC). This method has garnered growing interest among researchers in the field of UMS control due to its numerous advantages. It has been widely verified in the literature that sliding mode control-based approaches are able to maintain stable and satisfactory performance under a wide range of uncertain conditions, variations, and disturbances. Moreover, improved

Control of Underactuated Mechanical Systems. https://doi.org/10.1016/B978-0-44-324020-1.00011-7

performance of first-order SMC is verified by continuously adjusting the control parameters of the latter [4,5,11,12]. The advantages of time-varying sliding mode approaches are numerous, such as less manual gain tuning, preventing instability and adaptability of the control law to varying conditions. In this chapter, a broad overview of the conventional sliding mode approach is presented. Secondly, a brief recap of several adaptive sliding mode approaches is introduced. Then, a control solution is presented, for a specific class I, according to the classfication proposed by Olfati-Saber [10]. A particular case has been studied, namely the inertia wheel inverted pendulum. The proposed solution is based on an adaptation law that automatically estimates the switching SMC gain. Finally, numerical simulations and real-time experimental results are presented and discussed.

## 3.2 The conventional first-order SMC approach

This section provides a general survey of the first-order sliding mode control. The design of first-order SMC approach can be summarized in two steps. The first one consists in designing the sliding surface as a set of relationships between the system state variables. This means that the dynamics of the system confined to that sliding manifold will perform a desired behavior. The second one consists in designing a discontinuous control law that forces the system's state trajectories to reach the sliding surface and maintains them there, in spite of uncertainties/disturbances. The main advantage of such a controller is that the closed-loop response becomes insensitive to a certain uncertainty. This makes the approach an adequate candidate for robust control. Consider a nonlinear system as follows:

$$\dot{x} = f(x,t) + g(x,t)u \tag{3.1}$$

$$y = h(x,t) \tag{3.2}$$

where $x = (x_1 x_2, ..., x_n)^T \in \chi \subset \mathbb{R}^n$ denotes the state vector, $\chi$ is a bounded subset of $\mathbb{R}^n$, $f$ and $g$ are both uncertain nonlinear differentiable functions, $u \in U \subset \mathbb{R}$ is the input, and $U$ is a bounded subset of $\mathbb{R}$. Let $S$ be a set of switching surfaces defined by:

$$S = \{x \in \chi : s(x,t) = [s_1(x,t), ..., s_i(x,t), ..., s_m(x,t)]^T = 0\} \tag{3.3}$$

$s(x,t) = 0$ is called a dimensional sliding manifold of order $n - m$ in $\mathbb{R}^n$ and is represented by the intersection of $m(n-1)$ $s_i(x,t) = 0$ sliding manifold, where $m$ and $n$ denote, respectively, the number of control signals and number of system states. The motion on the set $S$ is known in the literature as a sliding mode. The main objective is to force the system trajectory to reach a switching surface through a discontinuous feedback law. The latter is described by:

$$u_i = \begin{cases} u_i^+ & \text{if } s_i(x) > 0 \\ u_i^- & \text{if } s_i(x) < 0 \end{cases} \tag{3.4}$$

where $u_i^+$ and $u_i^-$ are both continuous functions ($u_i^+ \neq u_i^-$) and $s_i(x)$ denotes the $i$th sliding surface. To let the closed-loop system reveal a sliding mode (or sliding regime) on the sliding manifold $s(x,t) = [s_1(x,t), ..., s_m(x,t)]^T = 0$, the next step is to verify the existence of an ideal sliding mode. In fact, the sliding surface must be locally attractive.

**Definition 3.2.1.** Consider the nonlinear system (3.1), a sliding motion or (an ideal sliding mode) exists on $S$ if there is a finite time $t_f$ such that the solution of (3.1) satisfies $s(x(t)) = 0$, for all $t \geq t_f$.

In this sense, the system states must approach the surface at least asymptotically. This can be explained geometrically that in the vicinity of the sliding surface, the tangent vector (or time derivative of the state vector) must point toward the sliding surface in the region of attraction:

$$lim_{s \to 0^+} \frac{\partial s}{\partial x}(f + gu) < 0 \quad \& \quad lim_{s \to 0^-} \frac{\partial s}{\partial x}(f + gu) > 0 \tag{3.5}$$

The existence of a sliding mode can be considered as a general stability problem. Accordingly, the Lyapunov method provides an effective framework for the stability analysis of the closed-loop system. This method requires finding a definite and generalized positive Lyapunov function with a negative time derivative in the region of attraction. However, there is no standard method for finding Lyapunov functions for arbitrary nonlinear systems. It is worth noting that for various systems, a convenient and appropriate positive Lyapunov function is often selected as:

$$V(x) = \frac{1}{2}s^2 > 0 \tag{3.6}$$

Due to its positivity, (3.6) ensures that the Lyapunov function is always positive for any non-zero value of the sliding surface $s$. It is a common useful choice to analyze the stability of linear systems. It is a convex function having a single minimum point. The first derivative of (3.6) is expressed as:

$$\dot{V}(x) = s\dot{s} \leq \eta|s| < 0 \tag{3.7}$$

This expression is called the $\eta$-reachability condition. It ensures that the sliding manifold is reached asymptotically in a finite time. Since differential equations represented by (3.1) and (3.4) involve discontinuous terms (due to the discontinuity of the control input), the solutions are to be understood in the

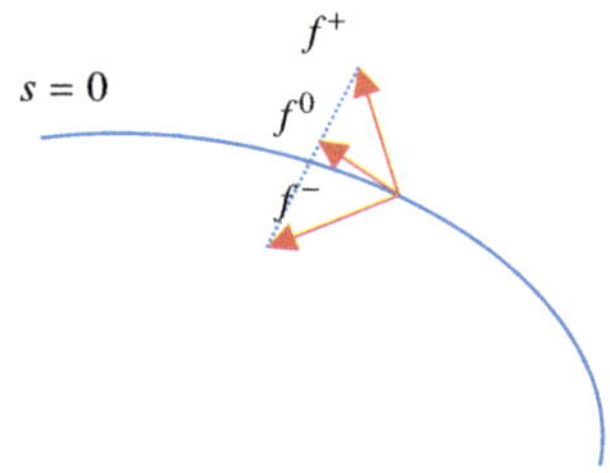


**FIGURE 3.1** Illustration of Filippov's solutions in sliding mode.

sense of Fillipov [1]. This means that the system trajectories of $\dot{x} = f(x, t, u)$ with the discontinuous control input (3.4) are the solution of:

$$\dot{x} = \xi f(x, t, u^+) + (1 - \xi) f(x, t, u^-) = f^0 \tag{3.8}$$

where $0 \leq \xi \leq 1$, $f^0$ denotes the velocity of the state trajectory and may be geometrically explained in the sense that the average dynamic of $f^0$ is tangent to the surface $s(x) = 0$, as illustrated in Fig. 3.1.

The authors in [2] proposed the technique of the continuous equivalent control. This method entails setting the expression of the first derivative of the sliding surface to zero.

The derivative of $s(x, t)$ with respect to time is obtained by:

$$\dot{s}(x, t) = \frac{\partial s}{\partial x} f(x) + \frac{\partial s}{\partial x} g(x) u_e q(t) \tag{3.9}$$

Assuming that $\dot{s}(x) = 0$, the related equivalent control of system (3.1) can be expressed by:

$$u_{eq} = -[\frac{\partial s}{\partial x} g(x)]^{-1} \frac{\partial s}{\partial x} f(x) \tag{3.10}$$

By substituting (3.10) into (3.1), we obtain the resulting dynamics as follows:

$$\dot{x}_r = f(x_r) - g(x_r)[\frac{\partial s}{\partial x_r} g(x_r)]^{-1} \frac{\partial s}{\partial x_r} f(x_r) \tag{3.11}$$

$$\dot{x}_r = (I - g(x_r)[\frac{\partial s}{\partial x_r} g(x_r)]^{-1} \frac{\partial s}{\partial x_r}) f(x_r) \tag{3.12}$$

**Example:**

Let us consider a simple pendulum with one degree of freedom, as illustrated in Fig. 3.2.

The equation of motion of a simple pendulum is expressed as:

$$ml^2 \ddot{\theta} + mgl \sin\theta = \tau \tag{3.13}$$

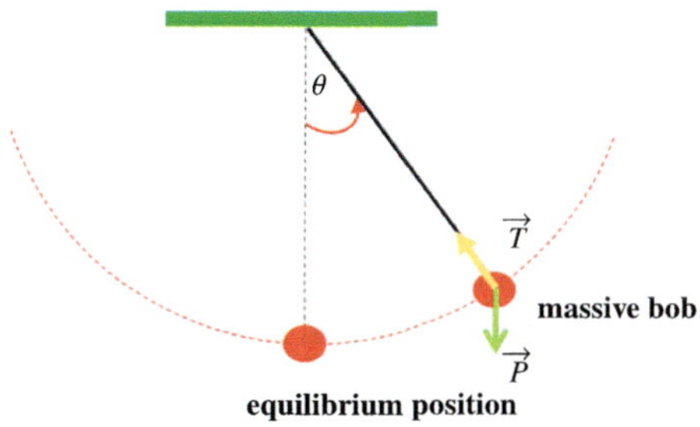


**FIGURE 3.2** Simple pendulum.

where $\theta$ is the angular position of the pendulum. $m$ denotes the mass of the pendulum, $l$ is its corresponding length. $g$ is the force of gravity. $\tau$ represents the control input. To represent the state space model of the pendulum, let us consider $x_1 = \theta$ $x_2 = \dot{\theta}$, which yields:

$$\dot{x}_1 = x_2 \tag{3.14}$$

$$\dot{x}_2 = \frac{g}{l}\sin x_1 + \frac{1}{ml^2}\tau \tag{3.15}$$

Let us consider the sliding surface $s = \rho x_1 + x_2$ as a linear relationship between the states, its first derivative will then be described as follows:

$$\dot{s} = \rho\dot{x}_1 + \dot{x}_2 \tag{3.16}$$

$$\dot{s} = \rho x_2 - \frac{g}{l}\sin x_1 + \frac{1}{ml^2}\tau \tag{3.17}$$

The expression of the equivalent control is obtained by considering $\dot{s} = 0$ as:

$$u_{eq} = ml^2(-\rho x_2 + \frac{g}{l}\sin x_1) \tag{3.18}$$

In this section, we have provided the basic concepts of the first-order SMC, the interested reader is referred to [3] for further details.

## 3.3 Adaptive sliding mode control for nonlinear systems

This section gives a brief recap of the adaptive sliding mode controllers proposed in the literature.

Huang et al. proposed an adaptive sliding mode controller (ASMC) in [4]. This control method guarantees the tracking performance of the system, without the knowledge of the system uncertainties upper bound. As discussed in [5], the main drawback of such an approach is that the control gain increases from the establishment of the sliding mode. It is known that there exists a proportional relation between the magnitude of chattering and the switching gain of the sliding mode controller. Thus this approach is not relevant especially for the real-time application, because it may result in a large oscillation of the control input signal (chattering). This is primarily the result of the controller's gain being overestimated with respect to uncertainties.

The authors in [6] proposed various control methodologies, to reduce the magnitude of chattering due to the presence of unmodeled dynamics. The first solution is based on an observer included in an auxiliary control loop scheme. In this case, the controller is mainly based on estimated states instead of measured states. The author also proposed another sliding mode control scheme based on a time-varying switching gain. The main characteristic of this solution is that the switching gain is decreasing, resulting in an attenuation of the chattering. The third control solution is based on a switching gain that depends on the continuous equivalent control.

Another adaptive switching gain was recently proposed in [7]. The control strategy offers the benefits of compensating for unknown external disturbances and variable parameters, and may also reduce the chattering effect. However, efficiency is only verified by simulation and a certain limitation may occur because of the complexity of the calculation, as mentioned by the author. In this chapter, we turn our attention to the adaptive control proposed in [5]. The authors proposed an adaptive sliding mode approach that guarantees a real sliding mode and allows us to obtain an adequate gain in relation to the magnitude of variation/uncertainties.

Consider an uncertain nonlinear system described by:

$$\dot{x} = f(x) + g(x)u \tag{3.19}$$

where $x \in X \sqsubset \mathbb{R}^n$ represents the state vector of the system. $u \in \mathbb{R}$: denotes the control input. $f, g \neq 0$: are bounded uncertain functions for $x \in X$.

**Theorem 3.3.1.** *Let $\Gamma_m$, $\Omega_m$, $\Omega_M$ be unknown bounds. Consider a sliding surface $s(x,t)$, its first derivative is expressed by:*

$$\begin{aligned} \dot{s} &= \frac{\partial s}{\partial t}\dot{x} + \frac{\partial s}{\partial x} f(x) + \frac{\partial s}{\partial x} g(x)u && (3.20) \\ &= \Gamma(x,t) + \Omega(x,t)u && (3.21) \end{aligned}$$

*where*

- $\Gamma(x,t) = \frac{\partial s}{\partial t}\dot{x} + \frac{\partial s}{\partial x} f(x)$;
- $\Omega(x,t) = \frac{\partial s}{\partial x} g(x)$;
- $|\Gamma(x,t)| \leq \Gamma_M$;
- $0 < \Omega_m \leq \Omega \leq \Omega_M$.

Consider the system (3.19) with the sliding dynamics described by (3.20) and the control input is described by the adaptive sliding mode control law

(3.22), hence, a real sliding mode (i.e., $S = \{x \in X \subset \mathbb{R}^n \| |s| < \delta\}$) is established for $t > t_{finite}$

$$u = -K(t)sign(s) \tag{3.22}$$

$$\dot{K} = \begin{cases} \bar{K}|s(x,t)|sign(|s(x,t)| - \xi) & \text{if } K(t) > \eta \\ \eta & \text{if } K(t) \leq \eta \end{cases} \tag{3.23}$$

where:

- $K(0)$, $\bar{K}$, and $\xi$ are positive constants;
- $\eta$ is a positive small constant that allows us to obtain a positive constant switching gain $K$;
- $K(t) \leq K^*$, where $K^*$ denotes a positive upper bound of the adaptive gain;
- $|s| < \delta$, with $\delta = \sqrt{\xi^2 + \frac{\Gamma_M^2}{\bar{K}\Omega_m}}$.

The major features of this approach are as follows:

- The bounds of uncertainties $\Gamma_M$, $\Omega_m$, and $\Omega_M$ exist but are not known.
- The obtained adaptive gain is not overestimated, but reduced to the lowest level, ensuring the establishment of the real sliding mode $|s| < \delta$.
- Referring to [5], the proof of stability analysis is verified by choosing the Lyapunov candidate function $V = \frac{1}{2}s^2 + \frac{1}{2\gamma}(K - K^*)^2$.

## 3.4 Proposed adaptive sliding mode control for class-I 2-Dof UMSs

The objective is to propose a revisited method to control a class of uncertain nonlinear underactuated systems, for which uncertainties are finite and bounded, but not known [5]. As some aspects of the UMS's behavior may be left uncontrolled or partially controlled, we propose to represent the dynamics model in a cascade structure of two sub-systems. It is the first step required to design the SMC for the 2-DOF underactuated mechanical system. As the system is decoupled, the control law gains to all uncontrollable degrees of freedom and makes the control design simpler, providing greater control flexibility and precision. This means that the control action can regulate each part of the system without significant coupling between them. Then, using new generalized configuration variables, the dynamics of the system can be rewritten into a cascaded nonlinear system into a hierarchical strict-feedback form [10]. All the transformation steps will be included in the following section. The general representation of an underactuated mechanical systems with two joints can be described in the following manner:

$$\begin{bmatrix} m_{11}(q) & m_{12}(q) \\ m_{21}(q) & m_{22}(q) \end{bmatrix} \begin{bmatrix} \ddot{q}_1 \\ \ddot{q}_2 \end{bmatrix} + \begin{bmatrix} h_1(q,\dot{q}) \\ h_2(q,\dot{q}) \end{bmatrix} = \begin{bmatrix} 0 \\ \tau \end{bmatrix} \tag{3.24}$$

where $q = (q_1, q_2) \in \mathbb{R}^{n-m} \times \mathbb{R}^m$, where $q_1$ represents the non-actuated configuration and $q_2$ is the actuated one. $m_{11}, m_{12}, m_{21}, m_{22}$ represent four elements of the inertia matrix, the vector $h_1(q, \dot{q}), h_2(q, \dot{q})$ regroups all the terms of Coriolis, centrifugal, and gravity terms, and $\tau \in \mathbb{R}$ denotes the control input. Given that the control action appears in the second equation of (3.24), the system belongs to the class I and can be transformed into another state representation known as the cascade strict-feedback form. It refers to a particular mathematical representation of an underactuated dynamic system. In this new representation, the system is organized hierarchically, where each level of control introduces additional dynamics. The strict-feedback structure ensures that control variables are well defined, and it simplifies the process of the control design.

**Definition 3.4.1.** If we consider the UMS model (3.24) with symmetry properties, then we can adopt a new coordinate representation as follows [8]:

$$z_1 = q_1 + \gamma(q_2) \tag{3.25}$$

$$\gamma(q_2) = \int_0^{q_2} m_{11}^{-1}(\theta) m_{12}(\theta) d\theta \tag{3.26}$$

$$z_2 = m_{11}(q_2) p_1 + m_{12}(q_2) p_2 \tag{3.27}$$

$$\xi_1 = q_2 \tag{3.28}$$

$$\xi_2 = p_2 \tag{3.29}$$

where $[z_1, z_2, \xi_1, \xi_2]^T$ now denotes the new state vector including the new decoupling configuration states. This change of coordinates is permitted only if the expression $m_{11}^{-1}(q_2) m_{12}(q_2)$ is an integrable term. From the above new configuration, the system dynamic model expressed by (3.24) can now be rewritten in the following model:

$$\dot{z}_1 = m_{11}^{-1}(\xi_1) z_2 \tag{3.30}$$

$$\dot{z}_2 = g_1(z_1 - \gamma(\xi_1), \xi_1) \tag{3.31}$$

$$\dot{\xi}_1 = \xi_2 \tag{3.32}$$

$$\dot{\xi}_2 = u \tag{3.33}$$

where $g_1(z_1, \xi_1) = \frac{d}{dt}\frac{\partial L}{\partial \dot{q}_1} = \frac{\partial K}{\partial q_1} - \frac{\partial V}{\partial q_1} = -\frac{\partial V(q)}{\partial q_1}$. Both $K$ and $V$ denote, respectively, the invariant kinetic energy with respect to $q_1$ and the whole potential energy of the system. The difference $L = K - V$ represents the Lagrangian equations and $u$ is the new scalar or virtual control input of the second subsystem presented in (3.30). The system is namely strict-feedback as $\dot{z}_1, \dot{z}_2, \dot{\xi}_2$ only depend on states $z_1, z_2, \xi_1, \xi_2$. This standard model representation enables us to develop the suggested control method for the stabilization of the second-order UMS with a single unactuated joint.

## 3.5 Design of sliding mode controller

The IWIP is classified as a flat system with kinetic symmetry with a dynamics including a constant inertia matrix. Please see Chapter 2 for a detailed system description. We take into account the following observations in pursuit of simplifying the dynamic model and control process. In this study, we do not include the DC motor model attached to the wheel. The DC motor dynamics can be complex, and if their contribution to the system's behavior is negligible or well understood, it may be more practical to exclude them for the sake of simplicity. Moreover, the goal is to control specific variables and to achieve certain performance criteria that are not strongly influenced by the motor dynamics, it might be unnecessary to include the dynamic motor model. In accordance with the physical setup, the frictions on both actuated and unactuated joints are minimal. Therefore in the dynamic modeling of the IWIP, we neglect the frictional forces within the system. Moreover, as the angular position of the inertia wheel does not play a significant role in the Lagrange equations and the dynamics, it is not considered as a state variable.

Let us recall some steps in the design of the IWIP, for more details the reader is referred to Chapter 2. The system's Lagrangian formulation is represented as:

$$L = \frac{1}{2}(I_1\dot{\theta}_1^2 + i_2(\dot{\theta}_1 + \dot{\theta}_2)^2) - m_0 g\cos(\theta_1) \tag{3.34}$$

The dynamic model resulting from the application of Lagrange's method is described as follows [9]:

$$(I_1 + i_2)\ddot{\theta}_1 + i_2\ddot{\theta}_2 - m_0 g\sin(\theta_1) = 0 \tag{3.35}$$

$$i_2(\ddot{\theta}_1 + \ddot{\theta}_2) = \tau \tag{3.36}$$

where $\theta = [\theta_1, \dot{\theta}_1, \dot{\theta}_2]^T$ is the state vector. $\theta_1$ is the angular position of the pendulum. $\dot{\theta}_1$ is the velocity of the pendulum and $\dot{\theta}_2$ denotes the inertia wheel velocity, $[\ddot{\theta}_1, \ddot{\theta}_2]^T$ represents the vector of acceleration of both the pendulum and rotating wheel, the constant of gravitational acceleration is denoted by $g = 9.8$ m/s$^2$, and $\tau$ is the control input. The terms $I_1$ and $m_0$ are expressed, respectively, by $I_1 = m_1(l_1)^2 + m_2(l_2)^2 + i_1$ and $m_0 = m_1 l_1 + m_2 l_2$. The system dynamical parameters are summarized in Table 3.1.

The transformation of coordinates (3.25)–(3.29) presented in Definition 3.4.1, leads to the following equations [8]:

$$z_1 = (I_1 + i_2)\dot{\theta}_1 + i_2\dot{\theta}_2 \tag{3.37}$$

$$z_2 = \theta_1 \tag{3.38}$$

$$z_3 = \dot{\theta}_2 \tag{3.39}$$

**TABLE 3.1 Dynamical parameters of IWIP.**

| Notation | Description | Units |
|---|---|---|
| $i_1$ | Body Inertia | [kg m$^2$] |
| $i_2$ | Wheel Inertia | [kg m$^2$] |
| $l_1$ | Length from pendulum base to pendulum center of mass | [m] |
| $l_2$ | Length from pendulum base to wheel center of mass | [m] |
| $m_1$ | Body mass | [kg] |
| $m_2$ | Mass of the wheel | [kg] |

where $[z_1, z_2, z_3]^T$ now becomes the new state variable. The updated representation of the system is then expressed as follows:

$$\dot{z}_1 = \frac{\partial L}{\partial \theta_1} = m_0 g \sin(z_2) \tag{3.40}$$

$$\dot{z}_2 = \frac{z_1}{I_1 + i_2} - \frac{i_2 z_3}{I_1 + i_2} \tag{3.41}$$

$$\dot{z}_3 = u \tag{3.42}$$

We propose a Lyapunov function candidate defined as:

$$V(z_1) = \frac{1}{2} z_1^2 \tag{3.43}$$

When we differentiate Eq. (3.43), we obtain:

$$\dot{V}(z_1) = z_1 \dot{z}_1 \tag{3.44}$$

We propose two desired trajectories defined as follows:

$$\theta_{1d} = z_{2d} = -\alpha \tanh(\sigma z_1) \tag{3.45}$$

$$\theta_{2d} = z_{3d} = 0 \tag{3.46}$$

where $0 < \alpha < \frac{\pi}{2}$ and $\sigma > 0$ are considered two positive constant parameters. Replacing $\dot{z}_1$ by its expression in Eq. (3.40), we obtain:

$$\dot{V}(z_1) = z_1 m_0 g \sin(z_2) \tag{3.47}$$

Then, we substitute (3.45) into (3.47), yielding:

$$\dot{V}(z_1) = z_1 m_0 g \sin(-\alpha \tanh(\sigma z_1)) \tag{3.48}$$

Given the following conditions, we can analyze the sign of $sin(-\alpha \tanh(\sigma z_1)$:

- **Condition 1** ($-1 < \tanh(\sigma z_1) < 1$): The hyperbolic tangent function $\tanh(x)$ maps its argument to a value between $-1$ and 1 for all real numbers $x$. As a result, $-1 < \tanh(\sigma z_1) < 1$ implies that $\tanh(\sigma z_1)$ falls within the interval $(-1, 1)$.
- **Condition 2** ($0 < \alpha < \frac{\pi}{2}$): This condition ensures that $\alpha$ is a positive angle between 0 and 90°.
- **Condition 3** ($0 < -\alpha \tanh(\sigma z_1) < \frac{\pi}{2}$): This condition ensures that the product $-\alpha \tanh(\sigma z_1)$ is a positive value but less than $\frac{\pi}{2}$.
- $sgn(\sin(-\alpha \tanh(\sigma z_1)))$: The sine function $\sin(x)$ is positive for $0 < x < \pi$, and negative for $\pi < x < 2\pi$. However, since $-\alpha \tanh(\sigma z_1)$ is positive and less than $\frac{\pi}{2}$ (as per Condition 3), we know that $-\alpha \tanh(\sigma z_1)$ falls within the interval $(0, \pi)$. Therefore $\sin(-\alpha \tanh(\sigma z_1))$ is positive.
- $sgn(-\alpha \tanh(\sigma z_1))$: We already know that $-\alpha \tanh(\sigma z_1)$ is positive (as per Condition 3).

Since both terms are positive, their signs are the same. Therefore $sgn(\sin(-\alpha \tanh(\sigma z_1))) = sgn(-\alpha \tanh(\sigma z_1))$ holds true under the specified conditions. We can infer that the following criteria guarantee the stability of the $z_1$-sub-system.
if $z_1 > 0$, then $\dot{V}(z_1) < 0$;
if $z_1 = 0$, then $\dot{V}(z_1) = 0$;
if $z_1 < 0$, then $\dot{V}(z_1) < 0$.

The sliding surface is then expressed as follows:

$$s = (\theta_{1d} - \theta_1) + \alpha_1(\theta_{2d} - \theta_2) \tag{3.49}$$

Two terms are compromised in the first-order sliding mode control law. The first term represents the equivalent continuous control and the second term corresponds to the discontinuous control. The expression of the first-order sliding mode control input for the UMS is described as follows:

$$u = \beta - Ksign(s) \tag{3.50}$$

where $K$ denotes the positive switching gain. The width of the boundary layer around the sliding surface is determined by the switching gain parameter in SMC. By using this boundary layer, chattering, which is the high-frequency oscillation occurring when the system is exactly on the sliding surface, can be reduced. $\beta$ represents the expression of the equivalent continuous control law ensuring that the derivative of the sliding surface equals zero. In other words, it is responsible for driving the system state to the desired sliding surface and keeping it there. Considering two error surfaces expressed as follows:

$$e_1 = z_2 - z_{2d} \tag{3.51}$$

$$e_2 = z_3 - z_{3d} \tag{3.52}$$

The first time derivatives of both $e_1$ and $e_2$ are determined by substituting the expressions of $\dot{z}_2$ and $\dot{z}_3$ from (3.41) and (3.42), which leads to:

$$\dot{e}_1 = \frac{z_1}{I_1 + i_2} - \frac{i_2 z_3}{I_1 + i_2} - \dot{z}_{2d} \tag{3.53}$$

$$\dot{e}_2 = u - \dot{z}_{3d} \tag{3.54}$$

Let us now consider the expression of the time-varying surface $s$:

$$s = \lambda e_1 + e_2 \tag{3.55}$$

The first derivative of (3.55) is then determined as follows:

$$\dot{s} = \lambda \dot{e}_1 + \dot{e}_2 \tag{3.56}$$

$$\dot{s} = \lambda(\frac{z_1}{I_1 + i_2} - \frac{i_2 z_3}{I_1 + i_2} - \dot{z}_{2d}) + u - \dot{z}_{3d} \tag{3.57}$$

The expression of $\beta$ is as follows:

$$\beta = \dot{z}_{3d} - \lambda(\frac{z_1}{I_1 + i_2} - \frac{i_2 z_3}{I_1 + i_2} - \dot{z}_{2d}) \tag{3.58}$$

The role of the SMC based on strict-feedback transformation is the stabilization of the IWIP. However, it is generally agreed that chattering is a common issue related to this control strategy, especially in the case of real-time implementation. The primary concern is associated with control gain tuning. The switching gain should be set to a moderate level since high switching gain enhances the system's robustness and tracking accuracy, while increasing the chattering in the control input. To address the chattering issue, various control strategies are proposed in the literature, such as optimal sliding mode control [13], the adaptive fuzzy-based SMC approach [14], and reinforcement learning (RL) [15]. In this chapter, we propose to implement an adaptive control gain as in [5], that guarantees a sliding mode, designed without a priori information about the uncertainties/perturbation bound. Consider now the system (3.35) and (3.36), the sliding surface $s(x,t)$ described in (3.55) with the variable switching gain $K()$, the expression of the control law becomes as follows:

$$u = \beta - K()\tanh(s) \tag{3.59}$$

where $\beta$ denotes the same equivalent control law. The term $-K(t)tanh(s)$ compensates for uncertainties with $K(t)$ and is estimated through the proposed adaptive laws. Inspired by the adaptation law proposed in [5], we propose the following updated expression of $K(t)$ as:

$$\dot{K} = \begin{cases} \bar{K}|\lambda e_1 + e_2| tanh(|\lambda e_1 + e_2| - \xi) & \text{if } K(t) > \eta \\ \eta & \text{if } K(t) \leq \eta \end{cases} \tag{3.60}$$

where all parameters $\bar{K}$, $\xi$, and $\eta$ are positive gains, the initial value of $K(0) > 0$ is also positive, and the gain $\eta$ denotes a small positive constant, enabling the estimation of gain value $K(t)$ to be exclusively positive. In contrast to the approach proposed in [5], we propose to use the tanh instead of the $sign$ function. The $sign$ function provides a discontinuous behavior that maps its positive input between 1 and $-1$, except zero. In this case, the system state is forced to switch abruptly between two sides of the sliding surface. In this case, the control action can exhibit chattering that could potentially damage the system's actuator. The hyperbolic tangent (tanh) function provides a continuous and smooth sliding surface that can help in reducing the chattering associated with SMC. In this case, the generated control input tends to be smooth and continuous, leading to fewer oscillations in the control action, with fewer control transitions. The hyperbolic tangent function is also incorporated into the adaptation law to achieve a smoother estimated parameter, particularly when it approaches its minimal value $\eta$. To prevent the controller's gain $K$ from being overestimated, the role of (3.60) is reducing the variable gain $K(t)$ while $(|s(x,t)| < \xi)$. In the next section, the closed-loop stability is presented.

## 3.6 Closed-loop stability analysis

The Lyapunov-based approach for the stability analysis of the uncertain nonlinear system (3.19) controlled by the adaptive sliding mode control is presented as:

$$V = \frac{1}{2}s^2 + \frac{1}{2\gamma}(K - K^*)^2 \tag{3.61}$$

The first derivative of (3.61) is then described as follows:

$$\dot{V} = s\Gamma - s\Omega K sign(s) + \frac{1}{\gamma}(K - K^*)\bar{K}|s|sign(s - \xi) \tag{3.62}$$

Replacing the $\Gamma$ and $\Omega$ by their bounds, yields:

$$\dot{V} = |s|\Gamma_M - |s|\Omega_m K + \frac{1}{\gamma}(K - K^*)\bar{K}|s|sign(s - \xi) \tag{3.63}$$

Let us put and subtract the term $|s|\Omega_m K^*$ in the right-hand side of (3.63):

$$\dot{V} = |s|\Gamma_M - |s|\Omega_m K \underbrace{-|s|\Omega_m K^* + |s|\Omega_m K^*} + \frac{1}{\gamma}(K - K^*)\bar{K}|s|sign(s - \xi) \tag{3.64}$$

$$\dot{V} = |s|(\Gamma_M - \Omega_m K^*) - |s|\Omega_m(K - K^*) + \frac{1}{\gamma}(K - K^*)\bar{K}|s|sign(s - \xi) \tag{3.65}$$

Rearranging (3.65), one obtains:

$$\dot{V} = (\Gamma_M - \Omega_m K^*)|s| + ((K - K^*)\{-\Omega_m|s| + \frac{\bar{K}}{\gamma}|s|sign(s - \xi)\} \quad (3.66)$$

Consider a positive constant $\omega_k$, we add the term in $-\omega_k|K - K^*| + \omega_k|K - K^*|$ into (3.66), and we obtain:

$$\dot{V} = -(-\Gamma_M + \Omega_m K^*)|s| - \omega_k|K - K^*| - F \quad (3.67)$$

where $F = |K - K^*|(-\Omega_m|s| + \frac{\bar{K}}{\gamma}|s|sign(s - \xi) - \omega_k)$. The expression of (3.67) becomes as follows:

$$\dot{V} = -\omega_s|s| - \omega_k|K - K^*| - F \quad (3.68)$$

We multiply and divide the first term by $\sqrt{2}$ and the second term by $\sqrt{2\gamma}$, (3.68) is then expressed as:

$$\dot{V} = -\frac{\sqrt{2}}{\sqrt{2}}\omega_s|s| - \frac{\sqrt{2\gamma}}{\sqrt{2\gamma}}\omega_k|K - K^*| - F \quad (3.69)$$

$$\dot{V} \leq -min\{\omega_s\sqrt{2}, \omega_k\sqrt{2\gamma}\}(-\frac{|s|}{\sqrt{2}}\omega_s|s| + \frac{|K - K^*|}{\sqrt{2\gamma}}) - F \quad (3.70)$$

$$\dot{V} \leq -min\sqrt{2}\{\omega_s, \omega_k\sqrt{\gamma}V^{\frac{1}{2}} - F \quad (3.71)$$

To define the sign of $\dot{V}$, there are a couple of possible cases $|s| > \xi$ and $|s| < \xi$. Referring to [5], it was verified that $s$ converges to the domain $|s| < \xi$.

## 3.7 Numerical simulation results

The goal of this work consists in the implementation of the proposed adaptive control scheme and the standard SMC on the testbed of the IWIP and evaluation of the robustness of both controllers under different operating conditions. The block diagram of the proposed control solution is depicted in Fig. 3.3. To sum up, the system is characterized by an unactuated joint between the frame and the pendulum body and an actuated one between the body and the inertia wheel. In this study, the main objective is the stabilization of the system around its unstable equilibrium point. The simulation results of both first-order SMC and the proposed ASMC are presented and discussed. The control algorithms are implemented using MATLAB® software of Math-Works. The values of both physical parameters and control gains are listed in Table 3.2. Two main scenarios are conducted:

- Scenario 1: Rejection of external disturbances: We consider that the system is subject to two external perturbations. The first disturbance ($u_{ex1} = 2$ N m) is

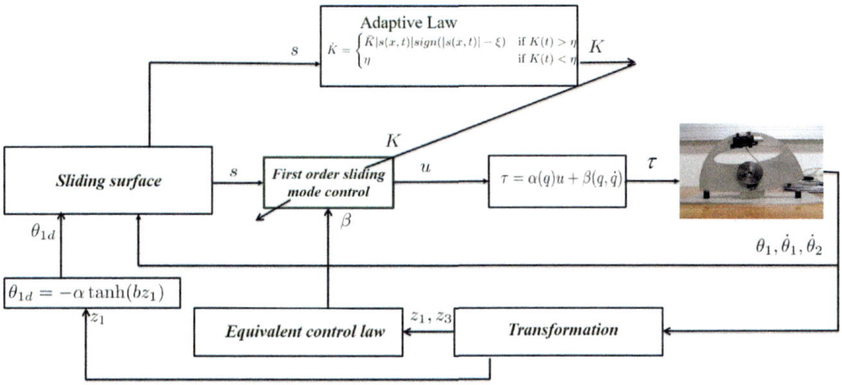


**FIGURE 3.3** The block diagram of the proposed adaptive sliding mode control based on strict-feedback transformation.

**TABLE 3.2** Simulation parameters.

| Notation | Value | Notation | Value |
|---|---|---|---|
| $i_1$ | $4.176e^{-4}$ [kg m$^2$] | $K$ | 1.6 |
| $i_2$ | 0.031468 [kg m$^2$] | $\bar{K}$ | 3.9 |
| $L_1$ | 0.06 [m] | $\eta$ | 0.3 |
| $l$ | 0.044 [m] | $\xi$ | 1.12 |
| $M$ | 3.30 [kg] | $\alpha_1$ | 0.1 |
| $m$ | 0.33 [kg] | $g$ | 9.81 [m s$^{-2}$] |

a punctual perturbation, added to the control at time instant ($t = 6$ s) and remained ($10^{-1}$ s). The second one ($u_{ex2} = 1$ N m) is a persistent perturbation introduced at time instant ($t = 12$ s) for (3 s).

- Scenario 2: Robustness towards parametric uncertainties: We consider the existence of uncertainty in the model parameter values. We have induced an uncertainty $\Delta I$ in the parameter $I = m_1(l_1)^2 + m_2(l_2)^2 + i_1$, as it represents a combination of multiple parameters.

### 3.7.1 Scenario 1: Rejection of external disturbances

External disturbances may have a significant impact on the closed-loop response, which represents a fundamental problem. In this section, the robustness of both controllers will be evaluated against external perturbations. Fig. 3.4 displays the obtained simulation results of the first scenario. The evolution of both external disturbances and control input are depicted in Fig. 3.4(**d**). It can be observed that the proposed approach is better than the standard sliding mode controller in terms of compensating both external disturbances. The control action swiftly stabilizes the system around its unstable equilibrium position. Note that the standard SMC approach is inappropriate when the angular position is

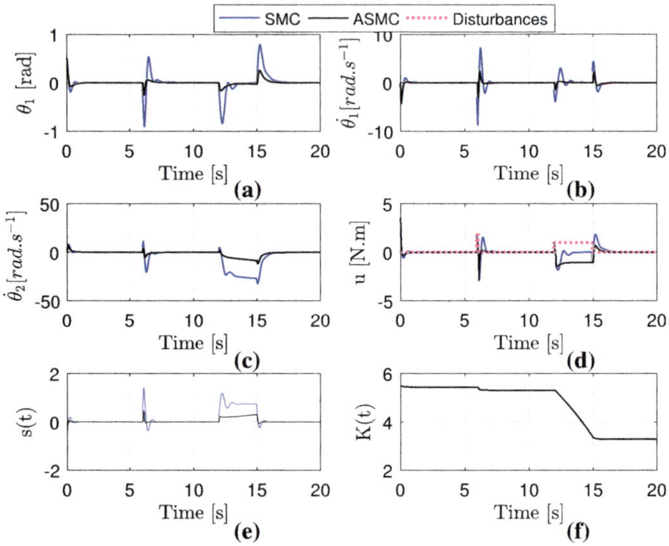


**FIGURE 3.4** Simulation results: scenario 1: External disturbance rejection: **(a)** Pendulum angular position, **(b)** Pendulum angular velocity, **(c)** Inertia wheel velocity, **(d)** Control input, **(e)** Sliding surface, **(f)** Estimated parameter.

constrained between [−0.5 rad, 0.5 rad]. It can be seen that the angular position at instants of time $t = 6$ s and $t = 15$ s, is far beyond the maximum value of the pendulum position (see Fig. 3.4(**a**)). The effectiveness of the proposed controller can also be observed in the evolution of both the angular velocities $\dot{\theta}_1, \dot{\theta}_2$ and the sliding surface.

### 3.7.2 Scenario 2: Robustness towards parametric uncertainties

We consider that the dynamics is subject to parametric uncertainties. The obtained simulation results of the second scenario are displayed in Fig. 3.5. In the case of $\Delta I = 50\% I$, we can see multiple oscillations, of which the amplitude increases over time, in the case of the SMC approach. This can cause system instability, although a convergence is observed in the case of the proposed approach. We can also note that even for a $\Delta I = 30\% I$, the SMC is not able to handle the parametric uncertainties. The results demonstrate the robustness of the proposed ASMC strategy toward parametric uncertainties. Additionally, it guarantees the non-overestimation of the switching control gain in both conducted scenarios, as shown in Fig. 3.4(**f**) and Fig. 3.5(**f**).

## 3.8 Real-time experimental results

Simulation is a great way to observe the behavior of various control approaches. It is more cost effective than real-time experiments. However, the complexity

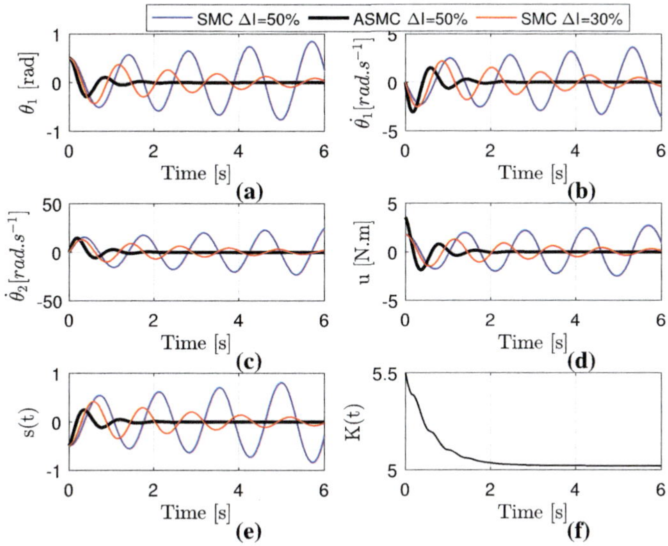


**FIGURE 3.5** Simulation results: scenario 2: robustness towards parametric uncertainties: **(a)** Pendulum angular position, **(b)** Pendulum angular velocity, **(c)** Inertia wheel velocity, **(d)** Control input, **(e)** Sliding surface, **(f)** Estimated parameter.

of the real system is more accurately replicated by experiments than by simulations. Even with a good identification, there is still a discrepancy between the dynamics of the experimental plant and the mathematical model. Moreover, real systems are more susceptible to parametric uncertainties and noise because of various factors, such as sensor noise, environment, or unmodeled dynamics. In this section, two conducted real-time case studies are discussed and explored in the following subsections. In the initial experiments, both controllers are tested in the nominal case. In the second, the system is subject to a significant external disturbance.

### 3.8.1 Scenario 1: Nominal case

A summary of the tuned parameters for both proposed control methods is presented in Table 3.3. Fig. 3.6 displays the obtained experimental results for the first experiment. In Fig. 3.6(**a**), we present the evolution of the angular position of the pendulum $\theta_1$. The evolution of the first derivative $\dot{\theta}_1$ is also depicted in Fig. 3.6(**b**). The evolution of the wheel velocity with respect to time $\dot{\theta}_2$ is presented in Fig. 3.6(**c**). The observed experimental results demonstrate that the proposed ASMC contributed to ensure better convergence than the conventional SMC. At the end of the experiment, we conclude that the proposed ASMC is faster with respect to time than the SMC. We observe additional oscillations around the equilibrium point in the evolution of position and velocity of the pendulum for the case of the conventional SMC. Additional time is required

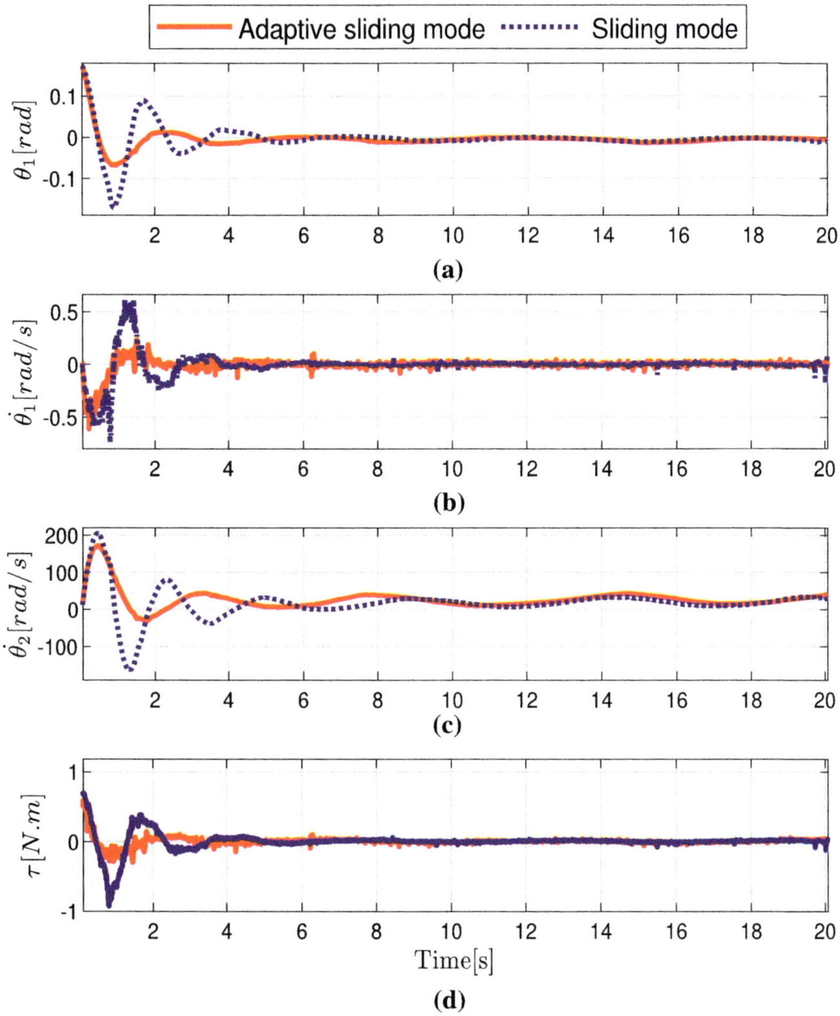


**FIGURE 3.6** Obtained experimental results for scenario 1: Nominal case. **(a)** Pendulum angular position, **(b)** Pendulum angular velocity, **(c)** Velocity of the inertia wheel, **(d)** Control input.

to achieve the stabilization for the case of SMC. Fig. 3.7 illustrates the time-dependent evolution of the specified sliding surface described in Eq. (3.49). Less varying responses are noted for the case of ASMC. After 0.42 s, the sliding surface of SMC is increased to a value of 0.25, while the sliding surface for ASMC did not reach 0.07 and remained close to zero. This improvement is attributed to the enhancement of the SMC through the proposed adaptation law. The phase diagram for both controllers is illustrated in Fig. 3.8. It can be seen that the pendulum is kept around its unstable equilibrium point, however, the convergence to the equilibrium point is ensured more rapidly for the proposed approach. The

**TABLE 3.3** Control gains used in real-time experiments.

| Parameter | Value | Parameter | Value |
|---|---|---|---|
| $\alpha_1$ | 0.33 | $\bar{K}$ | 16.9 |
| $\alpha$ | $\frac{\pi}{3}$ | $\xi$ | 5.6 |
| $K$ | 1.5 | $\eta$ | 2.05 |
| $\sigma$ | 1 | $K(0)$ | 3.9 |

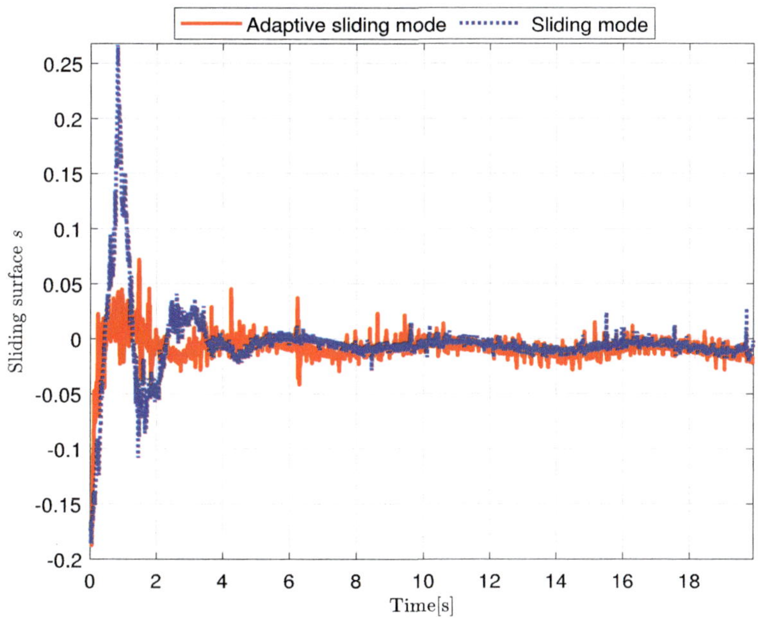


**FIGURE 3.7** Time evolution of the sliding surfaces in the nominal case.

SMC produces more oscillations of the pendulum around the equilibrium point. On the other hand, ASMC achieves stabilization rapidly. The phase diagram demonstrates the enhancement achieved by the adaptation law.

### 3.8.2 Scenario 2: External disturbances rejection

It is widely recognized that the presence of disturbances can lead to both instability and reduction of the controlled system performances. An external disturbance is applied to the experimental setup in the second experiment. At the start of the experiment, an additional mass, equal to $m_d = 0.34$ kg, is asymmetrically attached to the body of the pendulum. This attached mass will probably exert a continuous and persistent torque applied on the passive joint, due to its gravitational inertia. The performances of both controllers are analyzed and evaluated.

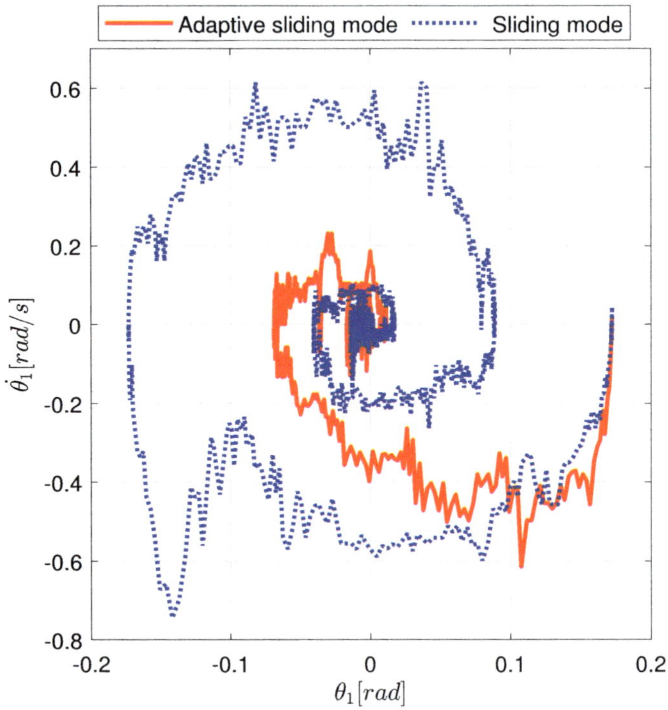


**FIGURE 3.8** The evolution of the phase diagram for both controllers in the nominal case.

Fig. 3.9 presents the evolution of both the angular positions and velocity evolution. Both controllers continue to oscillate around the unstable equilibrium point very close to zero; however, we can deduce from the beginning of the experiment that ASMC can promptly counteract the force applied by the mass. The amplitude of the velocity measured for ASMC is reduced with respect to the SMC. The phase trajectories confirmed the capacity of the control law adaptation to compensate for the effect of the applied disturbance. Fig. 3.9(**d**) presents the evolution of the control input for both controllers. High control magnitude is observed in the evolution of the control input in the case of the SMC. The amplitude of the first peak with a duration of $\Delta T = 1$ [s] is much closer to the maximum value tolerated by the actuator. In this case, the actuator may become saturated and, operating at its limits, lead to potential damage. On the other hand, the control input of the ASMC is gradually decreased to zero, as the pendulum approaches the vertical unstable equilibrium position. The phase diagram depicted in Fig. 3.10 presents the difference between both controllers in terms of convergence. We observe that the pendulum starts in a non-equilibrium position for both control approaches. The phase diagram illustrates the additional capacity of the adaptation law to improve the response of the conventional SMC.

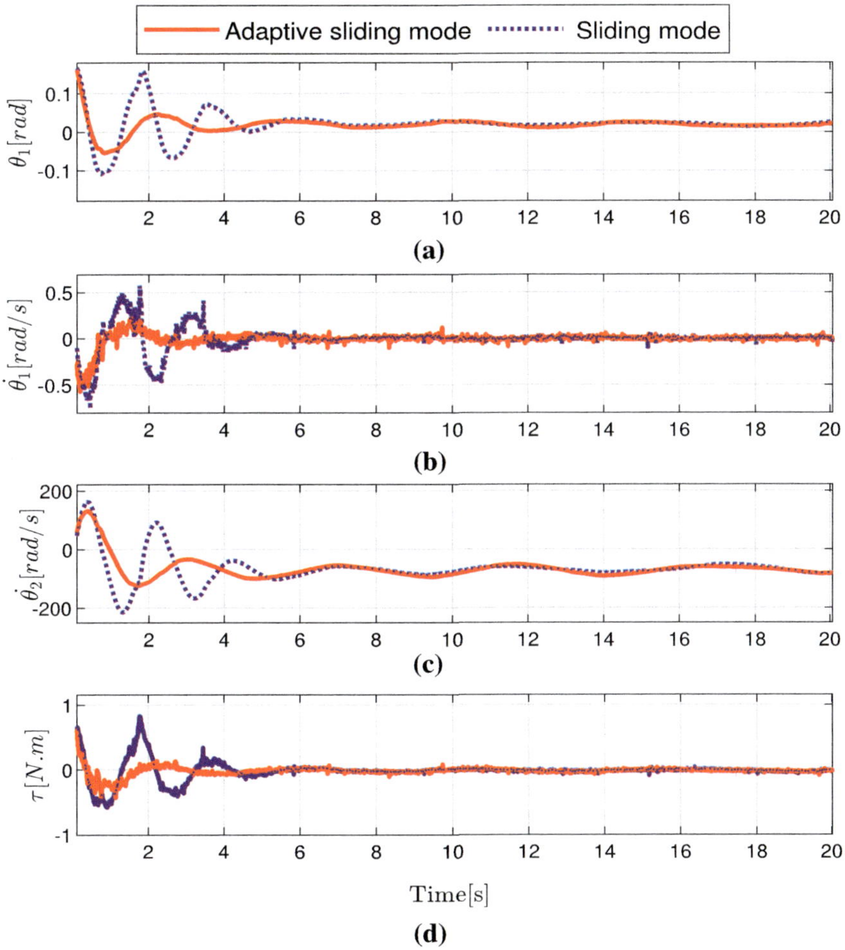


**FIGURE 3.9** Obtained experimental results for scenario 2: Rejection of persistent disturbances. **(a)** Pendulum angular position, **(b)** Pendulum angular velocity, **(c)** Velocity of the inertia wheel, **(d)** Control input.

To sum up, from the fact that the system is under the same disturbances for both controllers, we deduce that adjusting the gain control by the adaptation law can compensate for the external disturbance and results in a better stabilization of the pendulum around the unstable equilibrium point compared to the SMC. The evolution of estimated gain in the real-time experiment for both scenarios is presented in Fig. 3.11. It is observed in both scenarios that the estimated parameters are decreasing from the value $K(0) = 3.9$ until attaining a positive and very small bound very close to the tuned constant $\eta = 2.05$, which validates the real-time experiments. To evaluate the control algorithms' performance, we propose to compute the following criteria or indices in (3.72) and (3.73),

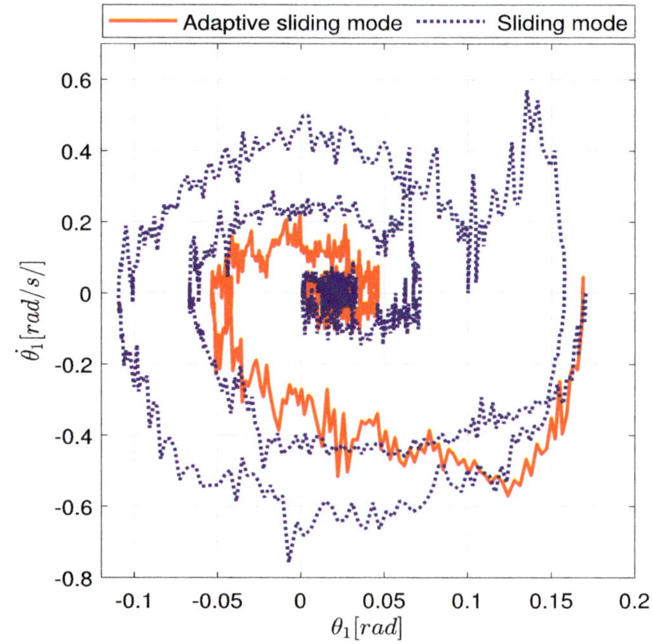


**FIGURE 3.10** Evolution of the phase diagram for scenario 2: Rejection of persistent disturbances.

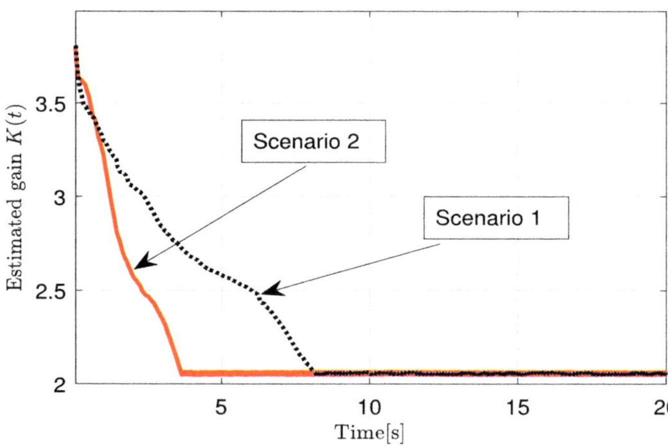


**FIGURE 3.11** The evolution of the estimated parameters versus time for both scenarios.

namely the integral square error (ISE) and integral absolute error (IAE). These performance metrics will provide insight into how well the proposed control system is tracking the reference trajectory:

$$ISE = \int e_{\theta_1}^2 dt \tag{3.72}$$

**TABLE 3.4** Control robustness evaluation of SMC and ASMC controllers.

| | IAE | ISE |
|---|---|---|
| Sliding mode | 0.8097 | 0.0379 |
| Adaptive sliding mode | 0.6374 | 0.0189 |
| Improvements | 21.27% | 50.13% |

$$IAE = \int |e_{\theta_1}| dt \tag{3.73}$$

where $e_{\theta_1}$ is chosen as the error of the passive joint $\theta_1$ at time t. Improvements of 21.27% and 50.13% are obtained on computing the ISE and IAE robustness metrics, see Table 3.4.

## 3.9 Conclusion

In this chapter, we propose a new approach based on an adaptive sliding mode control scheme in order to stabilize a two degrees of freedom underactuated inertia wheel inverted pendulum. The primary contribution lies first in the construction of a new sliding manifold based on both strict-feedback transformation and Lyapunov analysis. The proposed adaptation control law is then designed to estimate the control gain of the first-order sliding mode controller. A real-time implementation has been conducted on the inertia wheel inverted pendulum plant where the proposed approach has been compared with the conventional sliding mode control. Based on the experimental comparative analysis, we conclude that the proposed controller with variable control gain has significantly enhanced the performance of the system in terms of convergence and disturbance rejection compared to the first-order sliding mode with fixed switching gain. The chattering is still the major disadvantage of a first-order sliding mode controller, especially in real-time experiments. In the next chapter, we will extend the adaptation law with a robust integral of the sign error control for nonlinear uncertain underactuated mechanical systems.

## References

[1] A. Filippov, Differential Equations with Discontinuous Righthand Sides, Springer, Netherlands, 1988.

[2] V. Utkin, Sliding Modes in Control and Optimization, Springer-Verlag, Berlin, Heidelberg, 1992.

[3] J. Slotine, W. Li, Applied Nonlinear Control, Prentice-Hall, 1991.

[4] Y. Huang, T. Kuo, S. Chang, Adaptive sliding-mode control for nonlinear systems with uncertain parameters, IEEE Transactions on Systems, Man and Cybernetics. Part B. Cybernetics 38 (2008) 534–539.

[5] F. Plestan, Y. Shtessel, V. Brégeault, A. Poznyak, New methodologies for adaptive sliding mode control, International Journal of Control 83 (2010) 1907–1919.

[6] H. Lee, V. Utkin, Chattering suppression methods in sliding mode control systems, Annual Reviews in Control 31 (2007) 179–188.
[7] J. Guo, Application of a novel adaptive sliding mode control method to load frequency control, European Journal of Control 57 (2021) 172–178.
[8] R. Olfati-Saber, Global stabilization of a flat underactuated system: the inertia wheel pendulum, in: Proceedings of the 40th IEEE Conference on Decision and Control, 2001, pp. 3764–3765.
[9] A. Hfaiedh, A. Chemori, A. Abdelkrim, RISE controller for class I of underactuated mechanical systems: design and real-time experiments, in: The 3rd International Conference on Electromechanical Engineering (ICEE'2018), 2018.
[10] R. Olfati-Saber, Nonlinear Control of Underactuated Mechanical Systems with Application to Robotics and Aerospace Vehicles, Massachusetts Institute of Technology, 2001.
[11] V. Utkin, A. Poznyak, Adaptive sliding mode control with application to super-twist algorithm: equivalent control method, Automatica 49 (2013) 39–47.
[12] P. Li, Z. Zheng, Robust adaptive second-order sliding-mode control with fast transient performance, IET Control Theory & Applications 6 (2012) 305–312.
[13] M. Nikkhah, H. Ashrafiuon, K.R. Muske, Optimal sliding mode control for underactuated systems, in: 2006 American Control Conference, 2006, 6 pp.
[14] Y. Yang, Y. Yan, Attitude regulation for unmanned quadrotors using adaptive fuzzy gain-scheduling sliding mode control, Aerospace Science and Technology 54 (2016) 208–217.
[15] Q. Wang, A. Namiki, A. Asignacion Jr., Z. Li, S. Suzuki, Chattering reduction of sliding mode control for quadrotor UAVs based on reinforcement learning, Drones 7 (7) (2023) 420.

Chapter 4

# Nonlinear RISE feedback control scheme

## Contents



## 4.1 Introduction

The robust integral of the sign of the error (RISE) feedback control has been effectively implemented in fully actuated systems and has been verified as a promising control method. Indeed, in the literature, numerous studies have validated the effectiveness of RISE, in various domains [1–4], through simulations and experimental trials, showing good performance and stability under various conditions. Some applications include induction machines [5], tele-operation systems [2], aerial manipulators [1], parallel robots [6,7], underactuated mechanical systems [8], and exoskeletons [9–11], to name just a few.

The RISE controller generates a continuous control signal and it is characterized by the inclusion of an integral term in the control law that helps in avoiding issues like control chattering that are common in other robust control strategies like first-order Sliding Mode Control (SMC). It achieves asymptotic tracking error convergence and some studies prove that RISE can also achieve exponential stability [12,13].

The primary objective of this study is to introduce a novel redesign of the RISE feedback control for specific Class I UMSs [8]. The inertia wheel inverted

Control of Underactuated Mechanical Systems. https://doi.org/10.1016/B978-0-44-324020-1.00012-9 

pendulum belongs to Class I UMSs. It is regarded as an academic system to implement and explore new robust control approaches. The main control objectives of this system are the stabilization around its unstable equilibrium point, trajectory tracking, and the generation of stable periodic limit cycles. In this chapter, we will focus only on the problem of stabilization and improvement of robustness toward external perturbations and parameter variations.

## 4.2 Class I of underactuated mechanical systems

Due to their various properties and structures, UMSs are often dealt with on a case-by-case basis. Two known classifications were proposed in the literature. The first one is the classification due to [14]. The second one was proposed by [15]. The work of [14] is based on the graphical characterization of UMSs. This method is called a control flow diagram (CFD). It is a type of flowchart that allows us to represent the interaction forces that couples the degrees of freedom of an underactuated system. Three structures have been identified, namely chain structure, tree structure, and isolated point. Reza Olfati-Saber's main contribution consists of determining an explicit change of coordinates, based on the partial linearization of [16] transforms an UMS into the normal form. From the different obtained normal forms, we can categorize UMSs according to their structural properties and then design the appropriate control law for each class.

### 4.2.1 Noncollocated partial feedback linearization

The term "partial" indicates that not the entire dynamic system is linearized but only a part of it. This method simplifies the analysis and allows the use of well-established and understood linear control theory techniques. This section presents the noncollocated partial feedback linearization (NPFL) of an UMS. Consider an underactuated system expressed in the following form:

$$\begin{bmatrix} m_{00}(q) & m_{01}(q) & m_{02}(q) \\ m_{10}(q) & m_{11}(q) & m_{12}(q) \\ m_{20}(q) & m_{21}(q) & m_{22}(q) \end{bmatrix} \begin{pmatrix} \ddot{q}_0 \\ \ddot{q}_1 \\ \ddot{q}_2 \end{pmatrix} + \begin{bmatrix} N_0(q,\dot{q}) \\ N_1(q,\dot{q}) \\ N_2(q,\dot{q}) \end{bmatrix} = \begin{bmatrix} \tau_0 \\ \tau_1 \\ 0 \end{bmatrix} \tag{4.1}$$

with $q = (q_0, q_1, q_2) \in \mathbb{R}^{n-2m} \times \mathbb{R}^m \times \mathbb{R}^m$ the generalized coordinates. $\tau = (\tau_0, \tau_1)$. There is a lack of control input in the third equation of (4.1). According to [16], the system cannot be completely linearized, but can be partially linearized.

**Proposition 4.2.1.** *Consider the set $U \in (\mathbb{R}^n / det(m_{21}(q) \neq 0))$. The dynamics of (4.1) can be partially linearized thanks to the change of input of the form $\tau = \alpha(q)u + \beta(q,\dot{q})$:*

$$\dot{q}_0 = p_0 \tag{4.2}$$

$$\dot{p}_0 = u_0 \tag{4.3}$$

$$\dot{q}_1 = p_1 \tag{4.4}$$
$$\dot{p}_1 = f_0(q, p) + g_0(q)u_0 + g_2(q)u_2 \tag{4.5}$$
$$\dot{q}_2 = p_2 \tag{4.6}$$
$$\dot{p}_2 = u_2 \tag{4.7}$$

with

$$f_0(q, p) = -m_{21}^{-1} N_2(q, p)$$
$$g_2(q, p) = -m_{21}^{-1}(q) m_{22}(q)$$
$$g_0(q, p) = -m_{21}^{-1} m_{20}(q).$$

As $m_{21}$ is an invertible matrix in the set $U$, this change of control gives:

$$\ddot{q}_0 = u_0 \tag{4.8}$$
$$\ddot{q}_1 = u_1 \tag{4.9}$$

From the third equation in (4.1), we obtain the expression of $\ddot{q}_1$:

$$\ddot{q}_1 = f_0(q, p) + g_0(q, p)\ddot{q}_0 + g_2(q, p)\ddot{q}_2 \tag{4.10}$$

After the change of control we obtain:

$$u_1 = f_0(q, p) + g_0(q, p)u_0 + g_2(q, p)u_2 \tag{4.11}$$

Thus the new control can be expressed as:

$$u_2 = \frac{1}{g_2(q, p)}(u_1 - f_0(q, p) - g_0(q, p)u_0) \tag{4.12}$$

### 4.2.2 Collocated partial feedback linearization

This section presents the collocated partial feedback linearization (CPFL) of an UMS. The Euler–Lagrange equations of motion for an underactuated system with $n$ configuration variables and $m$ control inputs, $m < n$, can be expressed in the form:

$$\frac{d}{dt}(\frac{\partial L}{\partial \dot{q}}) - \frac{\partial L}{\partial q} = F(q)\tau \tag{4.13}$$

where $F(q)$ is a full column rank non-square matrix of external forces and $\tau \in \mathbb{R}^m$. The general form of equations of motion of UMS can be expressed as follows:

$$M(q)\ddot{q} + C(q, \dot{q})\dot{q} + G(q) = F(q)\tau \tag{4.14}$$

Consider that $F(q) = [0, I_m]^T$ and the configuration variable is $q_1 \in \mathbb{R}^{n-m}$, which is the unactuated one and $q_2 \in \mathbb{R}^m$ that represents the actuated configuration variable. The dynamics of an underactuated system is expressed in the following form:

$$\begin{bmatrix} m_{11}(q) & m_{12}(q) \\ m_{21}(q) & m_{22}(q) \end{bmatrix} \begin{bmatrix} \ddot{q}_1 \\ \ddot{q}_2 \end{bmatrix} + \begin{bmatrix} h_1(q,\dot{q}) \\ h_2(q,\dot{q}) \end{bmatrix} = \begin{bmatrix} 0 \\ \tau \end{bmatrix} \tag{4.15}$$

All systems in the form (4.15) can be globally partially linearized [17] thanks to the change of control.

**Proposition 4.2.2.** *Consider a global invertible change of control of the form $\tau = \alpha(q)u + \beta(q,\dot{q})$. Through this change of control, the dynamics of (4.15) is globally partially linearized as follows:*

$$\begin{aligned} \dot{q}_1 &= p_1 \\ \dot{p}_1 &= f_0(q,p) + g_0(q)u \\ \dot{q}_2 &= p_2 \\ \dot{p}_2 &= u \end{aligned} \tag{4.16}$$

*where $x = [q_1, p_1, q_2, p_2]^T$ is the state vector, $f_0(q,p) = -m_{11}^{-1}(h_1(q,\dot{q})$ and $g_0(q) = -m_{11}^{-1}(q)m_{22}(q)$.*

**Remark 1.** After the partial linearization, the system is represented in the form of two sub-systems. The control input $u$ appears in the nonlinear $(q_1, p_1)$-sub-system and the linear $(q_2, p_2)$-sub-system.

Taking the expression of $\ddot{q}_1$ from the first equation of (4.15) we obtain:

$$\ddot{q}_1 = -m_{11}^{-1}(h_1(q,\dot{q}) - m_{12}(q)\ddot{q}_2) \tag{4.17}$$

Substituting (4.17) into the second equation of (4.15) we obtain:

$$(m_{22}(q) - m_{21}(q)m_{11}^{-1}(q)m_{12}(q))\ddot{q}_2 + h_2(q,\dot{q}) - m_{11}^{-1}h_1(q,\dot{q}) = \tau \tag{4.18}$$

The change of control $\ddot{q}_2 = u$ gives:

$$\alpha(q) = (m_{22}(q) - m_{21}(q)m_{11}^{-1}(q)m_{12}(q)) \tag{4.19}$$

$$\beta(q,\dot{q}) = h_2(q,\dot{q}) - m_{11}^{-1}h_1(q,\dot{q}) \tag{4.20}$$

### 4.2.3 Normal forms of underactuated mechanical systems

As the control input in (4.16) appears in both sub-systems, we can decouple them through a global change of coordinates.

**Definition 4.2.1.** All UMSs with configuration vector $q = [q_1, q_2]^T$ and characterized by a kinetic symmetry with respect to $q_1$ ($M(q) = M(q_2)$), are classified into two Classes (I and II). We say an UMS of two DOFs belongs to the Class-I underactuated system, if the configuration variable $q_1$ is actuated. We named it a Class-II UMS if the configuration variable $q_2$ is unactuated.

**Definition 4.2.2** (Strict-feedback (SF) form). Using a global change of coordinates, the dynamics of Class I of UMSs can be globally and explicitly rewritten into cascade nonlinear systems in SF form. The change of coordinates is expressed as:

$$\begin{aligned} z_1 &= q_1 + \gamma(q_2) \\ z_2 &= m_{11}(q_2)p_1 + m_{12}(q_2)p_2 \\ \xi_1 &= q_2 \\ \xi_2 &= p_2 \end{aligned} \tag{4.21}$$

where $\gamma(q_2) = \int_0^{q_2} m_{11}^{(-1)}(\theta)m_{12}(\theta)d\theta$ and $[z_1, z_2, \xi_1, \xi_2]^T$ represents the novel state vector. Based on the new states, we obtain a new representation of the system composed of a linear (double integrator) sub-system and a nonlinear core sub-system [15]. The new system representation is then described as follows:

$$\begin{aligned} \dot{z}_1 &= m_{11}^{-1}(\xi_1)z_2 \\ \dot{z}_2 &= g_1(z_1 - \gamma(\xi_1), \xi_1) \\ \dot{\xi}_1 &= \xi_2 \\ \dot{\xi}_2 &= u \end{aligned} \tag{4.22}$$

where $g_1(q_1, q_2) = -\frac{\partial V(q)}{dq_1}$, $V(q)$ denotes the total potential energy of the system and $u$ is the new control input from CPFL.

**Corollary 4.2.1.** *Using an explicit global change of coordinates, the systems known as the Acrobot, the Inertia-Wheel pendulum, and the TORA all belong to the Class-I underactuated systems and can be globally transformed into the SF form (4.22).*

## 4.3 RISE controller for SISO systems

A nonlinear SISO system is defined by a system that has one input variable and one output variable. In other words, a single measurable output is affected by a single control input where the relationship between both is not linear. Consider an uncertain SISO nonlinear systems with dynamics described in the following equation:

$$m(q)\dot{q} + f(q) = u(t) \tag{4.23}$$

where the system state is denoted by $q \in \mathbf{R}$. $\dot{q}$ is its first time derivative. $m(q) \in \mathbf{R}$ and $f(q) \in \mathbf{R}$ both denote two uncertain functions. $u(t) \in \mathbf{R}$ represents the control input. Generally, in a tracking control problem, the control action should ensure that the dynamic system described in (4.23) follows a desired trajectory as closely as possible over time. Let a desired trajectory denoted by $q_d(t) \in \mathbf{R}$ be the reference signal. To achieve asymptotic tracking of the reference trajectory signal $q_d(t)$ by a continuous control law, the following hypotheses and assumptions are essential [18].

1. The first refers to the boundedness assumption of the uncertain functions $m(q)$ in (4.23) being limited or restricted from below by a positive constant $\underline{m} \in \mathbf{R}$ and from above by a positive non-decreasing function $\overline{m}(q) \in \mathbf{R}$.
2. Moreover, both $m(q)$ and $f(q)$ are differentiable with respect to the $q(t)$ and their derivatives are bounded.
3. The desired reference trajectory to be tracked is continuously differentiable. Its derivatives are bounded.

The expression of the feedback RISE control law $u(t)$ is described by:

$$u(t) = (k_s + 1)e_2 - (k_s + 1)e_2(t_0) + \int k_s + 1)\alpha_2 e_2(s) + \beta sign(e_2(s))ds \tag{4.24}$$

where $k_s > 0$, $\beta > 0$ and $\alpha_2 > 0 \in \mathbf{R}^+$ denotes a tuned positive control gain. $t_0$ is the initial time that ensures that the control input is equal to zero at $t = t_0$. The term $e_2$ is an auxiliary tracking error often described by a combination of both position error and velocity error as $e_2 = \dot{e}_1 + \alpha_1 e_1$. To conduct a comprehensive stability analysis, the reader is referred to [18].

## 4.4 RISE control for class I

Robust integral of the sign of the error (RISE) feedback control was successfully applied to fully actuated systems, and was validated as a promising control technique [9,10]. The main motivation of this contribution is to propose a new redesign of the RISE feedback control for the specific Class I of UMSs [8]. The following hypotheses are essential to ensure the design of the proposed nonlinear RISE controller. Consider the dynamics of a nonlinear system described in the following form:

$$M(x, \dot{x}, ..., x^{n-1})x^n + F(x, \dot{x}, ..., x^{n-1}) = U(t) \tag{4.25}$$

where $(.)^{(i)}(t)$ denotes the $i$th time derivative of (.), $U$ represents the control input vector, $X = [x, \dot{x}, ..., x^{n-1}, x^n] \in \mathbb{R}^n$ is the state vector, $M$ and $F \in \mathbb{R}^n$ are uncertain nonlinear functions.

**Property 1:** The uncertain nonlinear function matrix $M$ satisfies the following inequality:

$$\underline{m}\|\xi\|^2 \leq \xi^T M(.)\xi \leq \overline{m}(q)\|\xi\|^2 \tag{4.26}$$

It is bounded from below by a positive constant $\underline{m}$ and from above by a non-decreasing positive function $\overline{m}(q)$.

**Property 2:** The two functions $M(.)$ and $F(.)$ of the nonlinear system (4.25) are second-order differentiable.

**Property 3:** The desired reference trajectory is continuously differentiable with respect to time, up to the $(n+2)$th derivative.

The expression of the RISE control law requires some auxiliary error signals, denoted by $e_i(t), i = 1, 2$ and defined in the following manner:

$$e_1 = q_d - q \tag{4.27}$$

$$\dot{e}_1 = \dot{q}_d - \dot{q} \tag{4.28}$$

$$e_2 = \dot{e}_1 + \alpha_1 e_1 \tag{4.29}$$

$$r = \dot{e}_2 + \alpha_2 e_2 \tag{4.30}$$

where the two constant gains $\alpha_1$ and $\alpha_2$ are positive. $q_d$ and $\dot{q}_d$ are the desired trajectory and its first-time derivative, respectively. Some systems are not equipped with sensors required to measure the acceleration forces due to their very high cost. Hence, this quantity can be derived by differentiating the velocity signal. However, the limitation of this method is its high sensitivity to noise. In fact, a low level of noise in the velocity signal can lead to an amplified noise in the estimated acceleration signal. For this reason, in this case, the auxiliary error signal $r(t)$ is not used in the control design since it requires a numerical differentiation through the computational algorithm to obtain the second derivative of the position signal that is foremost a noisy signal. One practical solution to attenuate the noise level in the velocity estimation is proposed in a forthcoming section. The expression of the RISE control law is expressed as follows [18] and [19]:

$$\begin{aligned} u &= (k_s + 1)e_2 - (k_s + 1)e_2(0) + \nu_f \\ \dot{\nu}_f &= (k_s + 1)\alpha_2 e_2 + \beta sign(e_2) \\ \nu_f(0) &= 0 \end{aligned} \tag{4.31}$$

where the two constant gains $k_s$ and $\beta$ are positive and the $sign$ denotes the mathematical Signum function described as

$$sign(x) = \begin{cases} 0 & \text{if } x = 0 \\ \frac{x}{|x|} & \text{if } x \neq 0 \end{cases}$$

## 4.5 Closed-loop stability analysis

By substituting the expression $\ddot{\theta}_2$ in the first equation of (4.15), we obtain the same form as in (4.25) $m(q)\ddot{\theta}_1 + f(\theta_1) = u$, where $m(q) = I_0 + i_2$ and $f(\theta_1) = -m_0 g sin(\theta_1)$. When using all properties (1–3) and using a similar line of reasoning as in [18], the stability of the RISE controller can be demonstrated. Let the desired joint positions, velocities, and accelerations be represented by $q_d(t)$, $\dot{q}_d(t)$, and $\ddot{q}_d(t)$, respectively. Using the tracking error definitions from (4.27), the filtered tracking error $r(t)$ can be written as

$$r = \dot{e}_2 + \alpha_2 e_2 \tag{4.32}$$

where $\alpha_2 > 0$ denotes a positive design constant parameter. By multiplying the first derivative of (4.32) by $M(q)$, the open-loop error system is obtained:

$$m(q)\dot{r} = -\frac{1}{2}\dot{m}(q)r - e_2 - \dot{u} + N \tag{4.33}$$

In this case, the inertia matrix is constant, meaning $\dot{M}(q) = 0$, and thus the expression in (4.33) simplifies to:

$$m(q)\dot{r} = -e_2 - \dot{u} + N \tag{4.34}$$

where $N(.)$ represents a nonlinear and unmeasurable function, given by:

$$N(.) = m(q)\left(q_d^{(3)} + \alpha_2\dot{e}_2\right) + e_2 + \dot{f}(q) \tag{4.35}$$

Upon differentiating the control input with respect to time, we obtain:

$$\dot{u} = (k_s + 1)r + \beta sign(e_2) \tag{4.36}$$

Substituting (4.36) into (4.34) yields the following closed-loop system [8]:

$$m(q)\dot{r} = -e_2 - (k_s + 1)r - \beta sign(e_2) + N \tag{4.37}$$

Consider now an auxiliary function $N_d(.)$ expressed as follows:

$$N_d(.) = m(q)q_d^{(3)} + \dot{f}(q_d) \tag{4.38}$$

By adding and subtracting the function $N_d(t)$ from the right-hand side of the closed-loop error system (4.37), we obtain the following expression [8]:

$$m(q)\dot{r} = -e_2 - (k_s + 1)r - \beta sign(e_2) + \tilde{N} + N_d \tag{4.39}$$

where $\tilde{N}$ is defined by:

$$\tilde{N} := N - N_d \tag{4.40}$$

The function $N$, defined in (4.35), is continuously differentiable, and the function in (4.40) can be upper-bounded as follows:

$$\|\tilde{N}\| < \rho(\|z_\rho\|)\|z_\rho\| \tag{4.41}$$

where $\|z_\rho\|$ represents the Euclidean norm and $z_\rho(t)$ is defined as:

$$z_\rho := [e_1, e_2, r]^T \tag{4.42}$$

where $\rho : \mathbb{R}_{\geq 0} \to \mathbb{R}_{\geq 0}$ represents a non-decreasing and globally invertible function.

**Lemma 1.** *Let us consider the expression of the auxiliary function, $L(t) \in \mathbb{R}$ defined as:*

$$L = r(N_d - \beta sign(e_2)) \tag{4.43}$$

*If the control gain $\beta$ is chosen according to the following sufficient inequality expressed by:*

$$\beta > \|N_d(t)\|_{L_\infty} + \frac{1}{\alpha_2}\|\dot{N}_d(t)\|_{L_\infty} \tag{4.44}$$

*then*

$$\int_0^t L(\tau)d\tau \leq \zeta_b \tag{4.45}$$

*where $\| \|_{L_\infty}$ is the $L_\infty$ norm and $\zeta_b$ is a positive constant defined as*

$$\zeta_b = \beta|e_2(0)| - e_2(0)N_d(0) \tag{4.46}$$

**Theorem 1.** *The control law ensures that all system signals are bounded under closed-loop operation and*

$$\|e(t)\| \to 0 \quad as \quad t \to \infty \tag{4.47}$$

*provided that the control gain $\beta$ is selected according to (4.44), $\alpha_2 > \frac{1}{2}$ and the control gain $k_s$ is chosen sufficiently large relative to the system initial conditions.*

*Proof.* Let us consider the auxiliary function $P(t) \in \mathbb{R}$ expressed as follows:

$$P(t) = \zeta_b - \int_0^t L(\tau)d\tau \tag{4.48}$$

where $\zeta_b$ and $L$ are defined in (4.46) and (4.43), respectively. We define now the Lyapunov candidate $V : \mathbb{R}^{n+1} \times \mathbb{R}_{\geq 0} \longrightarrow \mathbb{R}_{\geq 0}$ as follows:

$$V(y,t) = \frac{1}{2}\sum_{j=1}^{n=2} e_j^2 + \frac{1}{2}mr^2 + P \tag{4.49}$$

where $y$ is defined as

$$y := [z_\rho^T \sqrt{P}]^T \tag{4.50}$$

and $z_\rho(t)$ is defined in (4.42). Since $m(q) > 0$ we can deduce that:

$$\underline{m} \leq m(q) \leq (\overline{m}(\|y\|)) \tag{4.51}$$

where $\underline{m}$ is a positive constant and $\overline{m}(q) : \mathbb{R}_{\geq 0} \longrightarrow \mathbb{R}_{\geq 0}$ is some non-decreasing function (see **Property 1**). Therefore we can bound (4.49) as follows:

$$\lambda_1\|y\| \leq V \leq \lambda_2(\|y\|)\|y\|^2 \tag{4.52}$$

where $\lambda_1 = (1/2)min\{1, \underline{m}\}$ and $\lambda_2(\|y\|) := max\{(1/2)\overline{m}(\|y\|), 1\}$.

After developing the time derivative of (4.49), we obtain:

$$\dot{V} = -\alpha_2 e_2^2 + e_1 e_2 - r^2 + r\tilde{N} - K_s r^2 + [r(N_d - Bsign(e_2)) - L] \quad (4.53)$$

where $\dot{P} = r(N_d - Bsign(e_2))$. By using the fact that $e_1 e_2 \geq (1/2)(e_1^2 + e_2^2)$ and (4.41) we can obtain an upper bound of (4.53) as follows:

$$\dot{V} \leq -\lambda_3 \|z_\rho\|^2 + |r|\rho(\|z_\rho\|)\|z_\rho\| - k_s r^2 \quad (4.54)$$

$$\dot{V} \leq -(\lambda_3 - \frac{\rho^2(\|z_\rho\|)}{4k_s})\|z_\rho\|^2$$

where $\alpha_2 > 1/2$ and $\lambda_3 = min\{(1/2), \alpha_2 - (1/2)\}$

We can conclude from (4.54) that

$$\dot{V} \leq -\gamma\|z_\rho\|^2 \quad (4.55)$$

with $k_s > \frac{1}{4\lambda_3}\rho^2(\|z_\rho\|)$ and $\gamma$ some positive constant. Based on Lemma 2 of [18], it can be verified that the closed-loop system states asymptotically converge to zero, thereby, $e_1 \to 0$, $e_2 \to 0$, $r \to 0$ as $t \to \infty$.

## 4.6 State estimation with robust Levant differentiator

In robotics and motion control systems, the presence of noise in measured signals is inevitable in any real-time experimental application. It certainly affects the performance of the system, especially when computing the first and second time derivatives of those noisy measured signals [8]. To overcome this problem, we propose to use the differentiator of [20], to provide the derivatives required to implement the RISE controller. The choice of the differentiator is motivated by its high precision and robustness in the presence of noise and parametric uncertainties [21]. The differentiator of [20] defined as a system of ordinary differential equations is described as follows:

$$\dot{w}_0 = v_0 \quad (4.56)$$

$$v_0 = w_0 - \lambda_2 L^{\frac{1}{3}}|w_0 - y|^{\frac{2}{3}} sign(w_0 - y) \quad (4.57)$$

$$\dot{w}_1 = v_1 \quad (4.58)$$

$$v_1 = w_1 - \lambda_1 L^{\frac{1}{2}}|w_1 - v_0|^{\frac{1}{2}} sign(w_1 - v_0) \quad (4.59)$$

$$\dot{w}_2 = -\lambda_0 L sign(w_2 - v_1) \quad (4.60)$$

If $t = 0$, $w_0 = y$, and $w_1(0) = w_2(0) = 0$. $y$ is the output of the system, $w_0$ is the estimated output. $w_1$ is its first time derivative and $w_2$ is its second time derivative. The gains $L$, $\lambda_0$, $\lambda_1$, and $\lambda_2$ are positive. The choice of a constant $L$ value can impact the overall behavior and characteristics of the system. In the equations for $v_0$ and $v_1$, the terms involving $L$ ($\lambda_2 L^{\frac{1}{3}}$ and $\lambda_1 L^{\frac{1}{2}}$) are associated

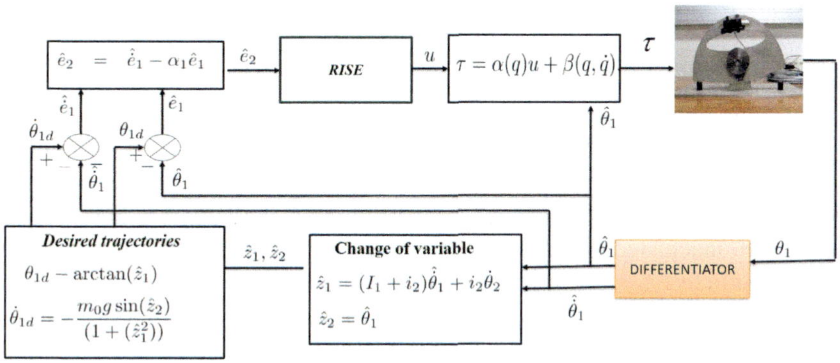


**FIGURE 4.1** Observer-based RISE control for inertia wheel inverted pendulum.

with absolute value terms $|w_0 - y|^{\frac{2}{3}}$ and $|w_1 - v_0|^{\frac{1}{2}}$, respectively. In this case, larger values of $L$ may amplify the impact of these absolute value terms. The behavior of the differentiator will also depend on the specific values chosen for the constants $\lambda_2, \lambda_1, \lambda_0$.

Based on a Levant differentiator, we propose to design the RISE controller for the inertia wheel inverted pendulum described in Chapter 2, using the global change of coordinates. The observer is employed to estimate the pendulum's angular velocity. In practice, calculating the velocity using the numerical derivative of the measured pendulum position amplifies the measurement noise. This noise affects the actuator's performance, potentially reducing the closed-loop system's effectiveness. Thus we propose estimating both the angular position and velocity of the pendulum with a differentiator to achieve smoother velocity and control input signals.

The block diagram of the proposed approach is depicted in Fig. 4.1. The dynamic model of the inertia wheel (described in Chapter 2) adheres to the first assumption of strict-feedback linearization, due to the constant nature of the inertia matrix. Consequently, the global coordinate transformation can be defined as follows:

$$z_1 = \frac{\partial L}{\partial \dot{\theta}_1} = (I_0 + i_2)\hat{\dot{\theta}}_1 + i_2\dot{\theta}_2 \tag{4.61}$$

$$z_2 = \hat{\theta}_1 \tag{4.62}$$

$$z_3 = \dot{\theta}_2 \tag{4.63}$$

where $Z = [z_1, z_2, z_3]$ represents the new state vector, and $\hat{\theta}_1, \hat{\dot{\theta}}_1$ are the estimated angular position and velocity of the pendulum, computed via the proposed Levant differentiator.

By applying the coordinate transformation, the system is reformulated into a strict-feedback structure as follows:

$$\dot{z}_1 = \frac{\partial L}{\partial \theta_1} = m_0 g \sin(z_2) \tag{4.64}$$

$$\dot{z}_2 = \frac{z_1}{I_0 + i_2} - \frac{i_2 z_3}{I_0 + i_2} \tag{4.65}$$

$$\dot{z}_3 = u \tag{4.66}$$

With this strict-feedback structure, the control input can influence both the actuated and unactuated coordinates. The resulting system comprises a cascade connection between a linear and a nonlinear sub-system, where $z_2$ serves as a virtual control input for the nonlinear $z_1$-sub-system. A Lyapunov-based analysis is used to ensure the global asymptotic stability of the nonlinear sub-system described by (4.64). The desired trajectory for the $z_1$-sub-system is assumed to follow a sigmoidal function, as expressed below:

$$\theta_{1d} = z_{2d} = -\arctan(z_1) \tag{4.67}$$

$$\dot{\theta}_{1d} = \frac{m_0 g \sin(z_2)}{1 + z_1^2} \tag{4.68}$$

A valid Lyapunov function is proposed as:

$$V(z_1) = \frac{z_1^2}{2} \tag{4.69}$$

Taking the time derivative of the Lyapunov candidate (4.69), we obtain:

$$\dot{V}(z_1) = z_1 \dot{z}_1 \tag{4.70}$$

Substituting $\dot{z}_1$ from (4.64) into (4.70) and incorporating the expression for (4.67), we obtain, after trigonometric simplifications:

$$\dot{V}(z_1) = z_1 m_0 g \sin(z_2) \tag{4.71}$$

$$\dot{V}(z_1) = z_1 m_0 g \sin(-\arctan(z_1)) \tag{4.72}$$

$$\dot{V}(z_1) = -z_1 m_0 g \frac{z_1}{\sqrt{1 + z_1^2}} \tag{4.73}$$

Based on the Lyapunov function (4.69), we can conclude that the desired sigmoidal function (4.67) globally stabilizes the nonlinear sub-system (4.64).

For $z_1 > 0$, $\dot{V}(z_1) < 0$.

For $z_1 = 0$, $\dot{V}(z_1) = 0$.

For $z_1 < 0$, $\dot{V}(z_1) < 0$.

The proposed desired trajectories can now be used for the design of the RISE controller. The control law in (4.31) is based on the estimated errors $\hat{e}_1$ and $\hat{e}_2$. Substituting (4.67) and (4.68) into (4.27), and replacing $\theta$ and $\dot{\theta}$ with their

**TABLE 4.1** Summary of the control design parameters.

| Parameter | Value | Parameter | Value |
|---|---|---|---|
| $\alpha_1$ | 15.3 | $I_0$ | 0.0130 |
| $\alpha_2$ | 1.5 | $m_0$ | 0.213 |
| $\beta$ | 14.05 | $K_s$ | 125.0 |
| $\Upsilon_0$ | 1.4800 | $\Theta_0$ | 0.0092 |

estimates $\hat{\theta}$ and $\hat{\dot{\theta}}$, yields:

$$\begin{aligned} \hat{e}_1 &= \hat{\theta}_1 - \theta_{1d} \\ \hat{\dot{e}}_1 &= \hat{\dot{\theta}}_1 - \dot{\theta}_{1d} \\ \hat{e}_2 &= \hat{\dot{e}}_1 - \alpha_1 \hat{e}_1 \end{aligned} \tag{4.74}$$

where $\theta_{1d}$ represents the desired trajectory, and $\dot{\theta}_{1d}$ is its first derivative. $\hat{\theta}$ and $\hat{\dot{\theta}}$ represent the estimated pendulum position and velocity, obtained through the proposed Levant differentiator.

Using equations (4.74) and (4.31), the control input based on the estimated states is expressed as:

$$u = (k_s + 1)\hat{e}_2 - (k_s + 1)\hat{e}_2(0) + \nu_f \tag{4.75}$$

$$\dot{\nu}_f = (k_s + 1)\alpha_2 \hat{e}_2 + \beta \,\text{sign}(\hat{e}_2) \tag{4.76}$$

## 4.7 Numerical simulation results

### 4.7.1 Scenario 1: Robustness towards parametric uncertainties

Generally, real systems exhibit uncertainties due to various factors, such as parameter variations and modeling inaccuracies. In this scenario, we test the controller under uncertain conditions. We consider an uncertainty term $\Delta I_0$ on the parameter $I_0$ and an uncertainty term $\Delta m_0$ on the parameter $m_0$ of the inertia wheel inverted pendulum. The obtained simulation results of the first scenario are presented and discussed. We selected the initial configuration to be $[\theta_1, \dot{\theta}_1, \dot{\theta}_2, \nu_f]^T = [0.2, 0, 0, 0]^T$. The control gains of the RISE controller are summarized in Table 4.1. Fig. 4.2 illustrates the evolution of the system states over time.

Case 1: $\Delta I_0 = 0\%$ and $\Delta m_0 = 0\%$: No uncertainties are introduced; hence, the simulation is conducted under nominal conditions. In the case of the RISE controller, the convergence of the pendulum angle, the pendulum velocity, and the inertia wheel velocity to the equilibrium point $[\theta_1, \dot{\theta}_1, \dot{\theta}_2] = [0, 0, 0]$ is clearly observable in a smaller time interval than the sliding mode control. The input torque converges to a steady-state value of zero, as depicted in Fig. 4.2.

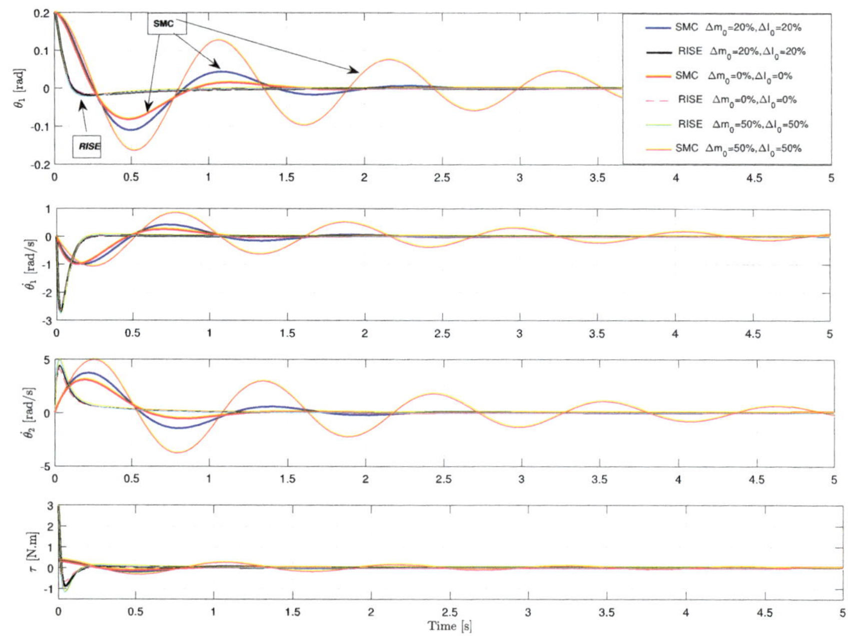


**FIGURE 4.2** Obtained simulation results for scenario 1: Robustness towards parametric uncertainties.

Case 2: $\Delta I_0 = 20\%$ and $\Delta m_0 = 20\%$: despite the introduced uncertainties, the proposed RISE control demonstrates the absence of oscillations, in contrast to the standard SMC. Notably, there is a substantial improvement of 79.2% and 83.2% in the evaluation of the ISE and IAE robustness indices, as detailed in Table 4.2.

Case 3: $\Delta I_0 = 50\%$ and $\Delta m_0 = 50\%$: By incorporating substantial variations of 50% in the model parameters $m_0$ and $I_0$, a notable distinction between both controllers becomes evident in the obtained simulation results. The introduced parametric uncertainties are not compensated by the sliding mode control, leading to increased oscillations around the equilibrium point. In contrast, the proposed RISE controller demonstrates its ability to effectively compensate for parametric uncertainties. It can be affirmed that the standard RISE controller proposed exhibits greater robustness in the face of parametric uncertainties compared to the SMC control.

### 4.7.2 Scenario 2: External disturbances rejection

The perturbation under consideration involves a non-sinusoidal sawtooth torque disturbance of magnitude [0.3 N m], applied to the pendulum at time $t = 4$ s, and lasting for a duration of 3 s. Subsequently, the second applied disturbance signal is described by $D = 0.1\sin(15t) + 0.15$, and it has been applied within

**TABLE 4.2** Control robustness evaluation of the SMC and RISE controllers for scenario 1.

| | IAE | ISE |
|---|---|---|
| SMC $\Delta I_0 = 20\%$, $\Delta m_0 = 20\%$ | 0.0965 | 0.0086 |
| RISE $\Delta I_0 = 20\%$, $\Delta m_0 = 20\%$ | 0.02 | 0.0014 |
| Improvement | 79.2% | 83.2% |

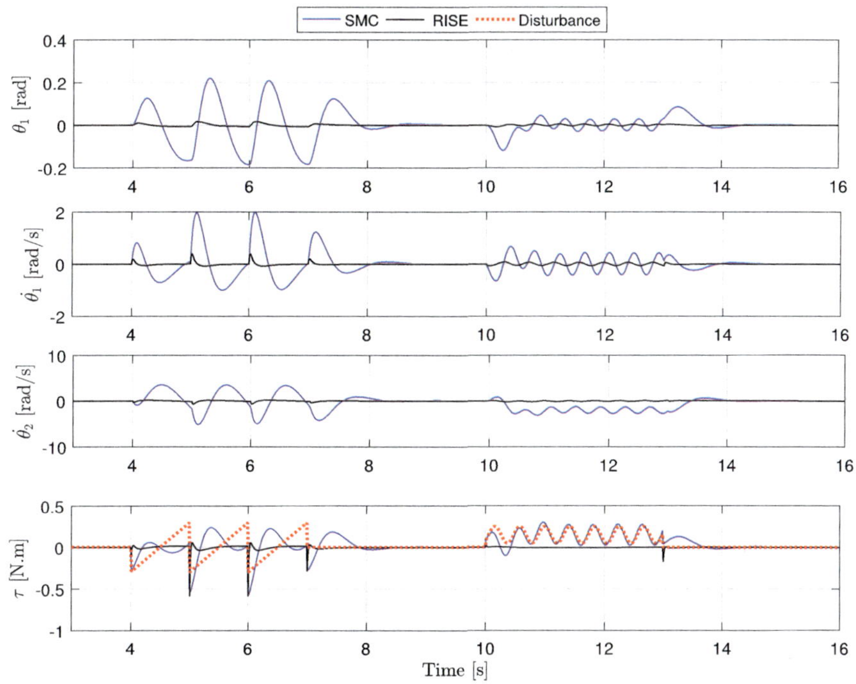


**FIGURE 4.3** Obtained simulation results for scenario 2: External disturbances rejection.

the time interval [10, 13] s. The simulation results are presented in Fig. 4.3, where the impact of the disturbances on the evolution of the system states is observable. The proposed approach responds to these disturbances, returning the system states to the desired equilibrium point. The control action effectively compensates for the deviation caused by the introduced external disturbances. The numerical simulation demonstrates the effectiveness of the proposed control approach, particularly in comparison to sliding mode control, especially when the system is subjected to the initial non-sinusoidal and periodic disturbance. As illustrated in Table 4.3, the performance of the proposed control approach is improved by up to 90%.

In the following section, we will analyze the performance of the proposed controller through real-time experiments.

**TABLE 4.3** Control performance evaluation of the SMC and RISE controllers for scenario 2.

| | IAE | ISE |
|---|---|---|
| SMC | 0.6228 | 0.0728 |
| RISE | 0.0533 | 0.0017 |
| Improvement | 91.4% | 97.7% |

**TABLE 4.4** Summary of the control design parameters.

| Parameter | Value | Parameter | Value |
|---|---|---|---|
| $\alpha_1$ | 7.3 | $\hat{\theta}_1(0)$ | 0.16 |
| $\alpha_2$ | 3.2 | $\lambda_0$ | 110 |
| $\beta$ | 14.05 | $\lambda_2$ | 50 |
| $K_s$ | 19.09 | $\lambda_1$ | 80 |

## 4.8 Real-time experimental results

Two experimental scenarios were conducted. In the first case, external disturbances were not taken into account, representing the experimental validation in the nominal case. The performance of the proposed observer-based RISE control, the standard RISE controller, and the sliding mode control are compared under nominal conditions. In the second scenario, the system is subject to persistent external disturbances. $RMSE = (\frac{1}{N}\sum_{i=1}^{N}(e_{\theta_1}(i))^2)^{\frac{1}{2}}$ is considered as a performance metric, where $N$ denotes the number of recorded samples, and $e_{\theta_1}$ represents the tracking error of the unactuated joint $\theta_1$. It is a valuable evaluation index as it offers a clear insight into the performance of the control algorithm and allows for a more comprehensive evaluation of the control strategies. The lower the RMSE value, the closer the system's response converges to the equilibrium point, indicating best and superior control performance. We also consider the input-torque-based criterion described by $E_\tau = \sum_{i=1}^{N}|\tau_{\theta_2}(i)|$ in order to analyze the performance of the system in terms of energy consumption, where $\tau_{\theta_2} = u$ represents the control input of the inertia wheel. The tuned experimental control design gains are listed in Table 4.4.

### 4.8.1 Scenario 1: Nominal case

In this part, we showcase the outcomes derived from three approaches, (i) the Sliding Mode Control (SMC), (ii) the RISE controller utilizing a numerical derivative, (iii) and the results obtained through the use of the Levant differentiator with the RISE controller. Fig. 4.4 presents the evolution of the angular position $\theta_1$ and velocity of the pendulum $\dot{\theta}_1$ with a zoomed-in view.

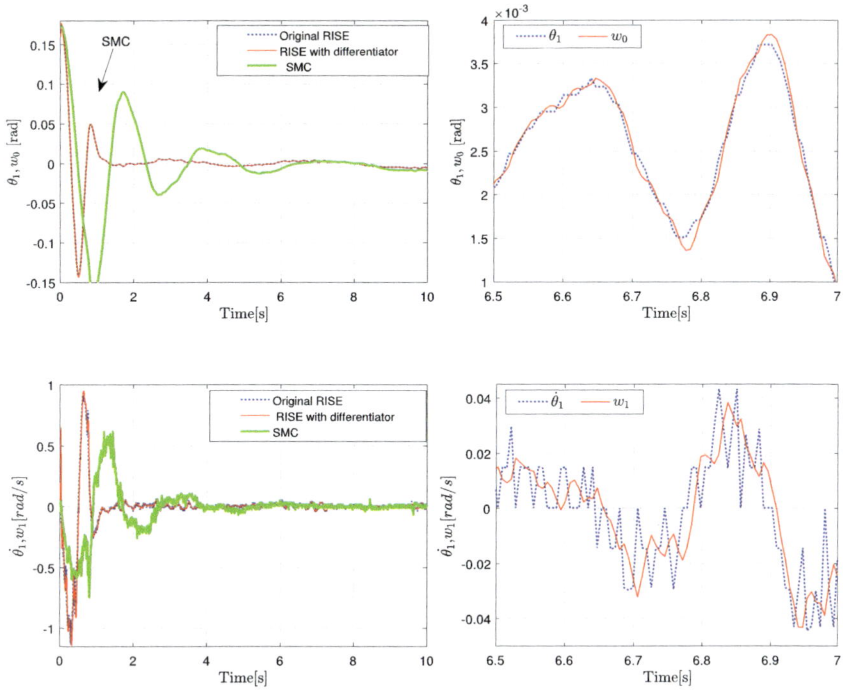


**FIGURE 4.4** Real-time experimental results for the first scenario: nominal case.

The real-time experimentally obtained results reveal that all system states converge towards the unstable equilibrium point. Notably, when compared to the first-order Sliding Mode Control, the response of both the pendulum and the wheel is faster in the case of the standard RISE controller and the observer based-RISE approach. For enhanced clarity, two zoom plots are provided in Fig. 4.4, focusing on the time intervals [6.5, 7] s. In these figures, the measured states of the standard RISE controller are depicted with dashed lines, while the estimated signals obtained from the differentiator are represented with solid lines. The stabilization of the pendulum around its unstable equilibrium is evident, showing an effective control strategies. However, a noticeable difference emerges in the zoomed plots between the numerical derivative of the pendulum position and its estimated counterpart generated by the differentiator. It can be noted that the utilization of the differentiator significantly attenuates the noise, presenting a significant improvement over the numerical derivative approach. Both control input and auxiliary error $e_2$ for the three controllers are displayed in Fig. 4.5.

The graphs are zoomed within the time interval [6.5, 10] s, where the continuous red line refers to the proposed control approach, and the blue dashed line corresponds to the outcomes of the standard RISE without the differentiator. The

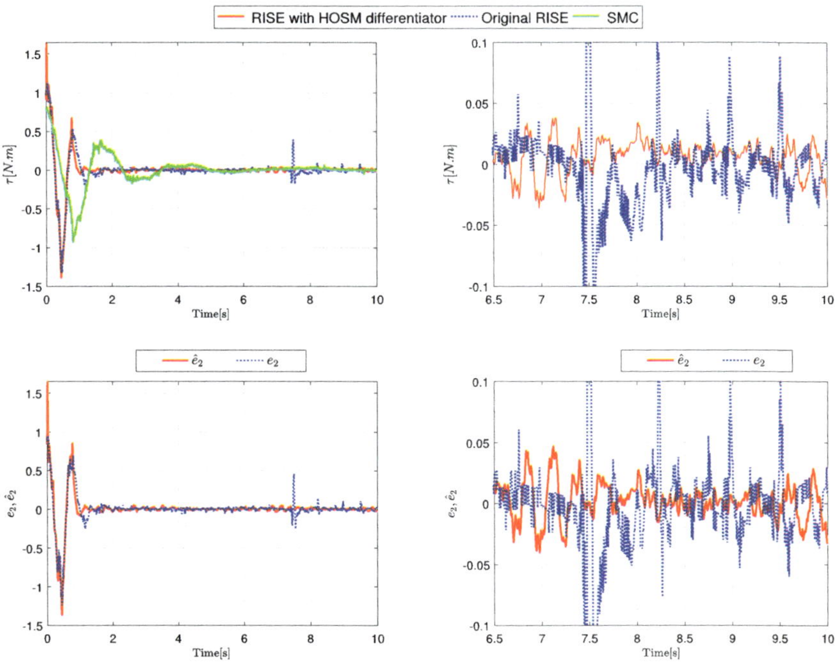


**FIGURE 4.5** Evolution of the control input and auxiliary error $e_2$ in real-time experiments: nominal case.

estimated auxiliary error $e_2$ is a smooth signal with respect to the measured one, although the standard RISE control ensures more effective stabilization of the system. The implementation of the Levant differentiator results in noise attenuation in the velocity of the pendulum, leading to a smoother control torque signal. Using a numerical derivative may tend to amplify any inherent noise in the measurement signals. Such noise may arise from interference by other electrical sources, hardware, sensors, or the environment. The real-time experimental results clearly demonstrate the superiority of both the original RISE and the proposed observer-based RISE approach compared to the SMC. Additionally, the use of Levant's differentiator provides enhanced real-time velocity estimation, outperforming the computed numerical derivation. Performance-evaluation metrics are computed to compare the effectiveness of the three approaches, see Table 4.5. The improvements of the RISE-based approach are computed with respect to the first-order sliding mode control. The best performance refers to the smallest value of the performance index and highest improvement. In summary, the obtained results show clearly that both the proposed observer-based RISE control scheme and the original RISE control outperform the SMC in terms of convergence and energy consumption. It can be concluded that satisfactory results are provided by a standard RISE controller without a differentiator. However, the proposed control approach has the best performance evaluation index

**TABLE 4.5 Scenario 1: Quantification of the performance achieved through various evaluation criteria.**

| Criteria | SMC | Standard RISE | Improvement | RISE with differentiator | Improvement |
|---|---|---|---|---|---|
| RMSE [rad] | 0.0051 | 0.0012 | 76.4% | $8.643e^{-4}$ | 82.35% |
| ISE | 0.0239 | 0.0098 | 59% | 0.0089 | 62.760% |
| IAE | 0.3961 | 0.165 | 58.34% | 0.1548 | 60.92% |
| $E_\tau$ [N m] | 189.8981 | 163.96 | 13.66% | 134.09 | 18.22% |

and improvement compared to the standard RISE approach by achieving better tracking accuracy and reduced control effort. This is attributed to the incorporation of the differentiator.

### 4.8.2 Scenario 2: Disturbances rejection

In this section, we test the ability of the proposed approach to mitigate and counteract external disturbances. Persistent disturbances rejection is a key aspect of control system design, particularly in real-time applications where external influences are continuous and may impact the system's behavior over an extended period of time. The goal in this scenario is to maintain the system's desired performance and stabilization around the unstable point under challenging and changing external conditions. An additional mass is attached to the pendulum body during the time interval [10.5, 16.5]. Zoom plots of the experimental results are depicted in Fig. 4.6. Both control approaches effectively compensate for persistent disturbances, leading to the convergence of the measured position and velocity of the pendulum to steady states. However, a closer examination of the zoomed plots reveals that the RISE control based on the differentiator yields superior performance compared to the RISE control. This improvement is particularly noteworthy in the estimation of the angular position and its derivative for the pendulum. The enhanced accuracy in estimating these states has influenced the control input signal, contributing to a more effective rejection of disturbances in the controlled system. In essence, the use of the differentiator with the RISE controller enhances the system's ability to counteract persistent disturbances, thereby improving overall performance. The results are also confirmed when comparing the performance through evaluation criteria, as stated in Table 4.6. The best obtained performance of the proposed approach is $5.853e^{-3}$ and 0.322 with an improvement of 36.83% and 12.47% for RMSE and IAE, respectively.

## 4.9 Conclusion

For the design of the RISE controller, the model of the system is first transformed into a strict-feedback form, with a Lyapunov-based analysis two new

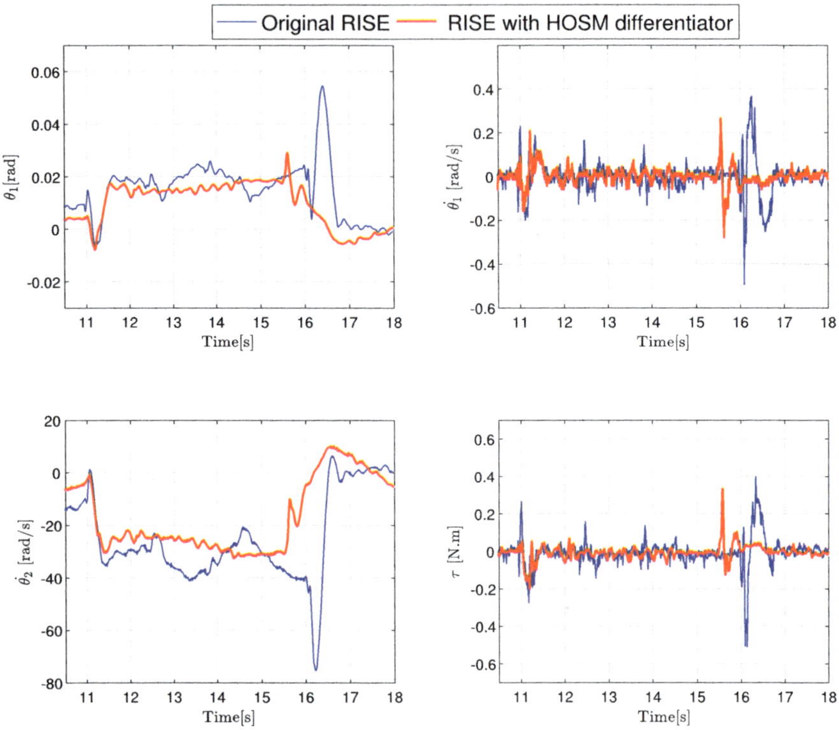


**FIGURE 4.6** Real-time experimental results for the second scenario: Persistent disturbance rejection.

**TABLE 4.6 Quantification of the performance through different evaluation criteria for scenario 2.**

| Criteria | Standard RISE | Proposed approach | Improvement |
|---|---|---|---|
| RMSE [rad] | $9.265e^{-3}$ | $5.853e^{-3}$ | 36.83% |
| ISE | $14.4905e^{-3}$ | $14.35766e^{-3}$ | 0.92% |
| IAE | $368.0434e^{-3}$ | $322.1330e^{-3}$ | 12.47% |
| $E_\tau$ [N m] | 218.7362 | 206.8575 | 5.43% |

desired trajectories are defined. Then, the stability analysis of the closed-loop system and simulation results for different operating conditions were addressed and show clearly that the proposed RISE control approach ensures a better stabilization and robustness against different parametric uncertainties and external disturbances compared to the SMC approach. However, the limitation of this approach was experimentally noted through the amplified noise, resulting from the numerical derivative action in the control input law. To overcome this drawback, a Levant differentiator was proposed and implemented to estimate the system states, including the first derivative of the angular position instead of the use of the numerical derivative of the noisy measured signal. Based on the obtained experimental results and further analysis based on some performance-evaluation indices, the use of a Levant differentiator with a RISE controller has significantly attenuated the effect of the noise included in the measured signals and improved the performance in terms of tracking error and energy consumption compared to both the original RISE and SMC approaches.

## References

[1] D. Lee, J. Byun, H. Jin Kim, RISE-based trajectory tracking control of an aerial manipulator under uncertainty, IEEE Control Systems Letters 6 (2022) 3379–3384.

[2] Y. Kawai, T. Namerikawa, Passivity and RISE based robust control for bilateral teleoperation system with communication delay, in: Proceedings of the IEEE International Conference on Control Applications, 2013, pp. 270–275.

[3] D. Li, Z. Xiang, Ye M. Distributed, Robust Nash equilibrium seeking for double-integrator-type players without velocity measurements, in: ASCC 2022–2022 13th Asian Control Conference, Proceedings, 2022, pp. 1905–1910.

[4] T. Nguyen, H. Nguyen, P. Dao, N. Vu, On robust control of permanent magnet synchronous generators using robust integral of error sign, Advances in Intelligent Systems and Computing 1284 (2021) 156–166.

[5] Y. Rkhissi-Kammoun, J. Ghommam, M. Boukhnifer, F. Mnif, RISE feedback control for induction machine in electric vehicle applications, in: Proceedings of 2015 7th International Conference on Modelling, Identification and Control, ICMIC 2015, 2016.

[6] G. Hassan, A. Chemori, M. Gouttefarde, M. El Rafei, C. Francis, P. Hervé, D. Sallé, A new augmented RISE feedback controller for pick-and-throw applications with PKMs, IFAC-PapersOnLine 55 (2022) 19–25.

[7] J. Escorcia-Hernandez, A. Chemori, H. Aguilar-Sierra, Adaptive RISE feedback control for robotized machining with PKMs: design and real-time experiments, IEEE Transactions on Control Systems Technology 31 (2023) 39–54.

[8] A. Hfaiedh, A. Chemori, A. Abdelkrim, Observer-based robust integral of the sign of the error control of class I of underactuated mechanical systems: theory and real-time experiments, Transactions of the Institute of Measurement and Control 44 (2022) 339–352.

[9] A. Hfaiedh, N. Khraief, S. Belghith, RISE control of patient-upper-limb-exoskeleton for passive rehabilitation, in: 2023 IEEE Third International Conference on Signal, Control and Communication (SCC), 2023, pp. 1–6.

[10] A. Hfaiedh, S. Bembli, N. Khraief, S. Belghith, Control of patient-upper-limb-exoskeleton: stability and robustness analysis, in: 2023 IEEE Third International Conference on Signal, Control and Communication (SCC), 2023, pp. 1–6.

[11] K. Sherwani, N. Kumar, A. Chemori, M. Khan, S. Mohammed, RISE-based adaptive control for EICoSI exoskeleton to assist knee joint mobility, Robotics and Autonomous Systems 124 (2020) 103354.

[12] O. Patil, K. Stubbs, P. Amy, W. Dixon, Exponential stability with RISE controllers for uncertain nonlinear systems with unknown time-varying state delays, in: Proceedings of the IEEE Conference on Decision and Control, 2022, pp. 6431–6435.

[13] O. Patil, A. Isaly, B. Xian, W. Dixon, Exponential stability with RISE controllers, IEEE Control Systems Letters 6 (2022) 1592–1597.

[14] D. Seto, J. Baillieul, Control problems in super-articulated mechanical systems, IEEE Transactions on Automatic Control 39 (1994) 2442–2453.

[15] R. Olfati-Saber, Nonlinear Control of Underactuated Mechanical Systems with Application to Robotics and Aerospace Vehicles, PhD thesis, Massachusetts Institute of Technology, Cambridge, MA, USA, 2001.

[16] M. Spong, Partial feedback linearization of underactuated mechanical systems, in: Proceedings of IEEE/RSJ International Conference on Intelligent Robots and Systems (IROS'94), vol. 1, 1994, pp. 314–321.
[17] M. Spong, Energy based control of a class of underactuated mechanical systems, IFAC Proceedings Volumes 29 (1996) 2828–2832, in: 13th World Congress of IFAC, San Francisco USA, 30 June – 5 July, 1996.
[18] B. Xian, D. Dawson, M. Queiroz, J. Chen, A continuous asymptotic tracking control strategy for uncertain nonlinear systems, IEEE Transactions on Automatic Control 49 (2004) 1206–1211.
[19] N. Fischer, D. Hughes, P. Walters, E. Schwartz, W. Dixon, Nonlinear RISE-based control of an autonomous underwater vehicle, IEEE Transactions on Robotics 30 (2014) 845–852.
[20] A. Levant, Higher-order sliding modes, differentiation and output-feedback control, International Journal of Control 76 (2003) 924–941.
[21] H. Imine, L. Fridman, H. Shraim, M. Djemai, Observation and identification via HOSM-observers, in: Sliding Mode Based Analysis and Identification of Vehicle Dynamics, 2011, pp. 5–24.

Chapter 5

# Model reference adaptive IDA-PBC approach

**Contents**



## 5.1 Introduction

Research in the field of controlling underactuated mechanical systems (UMSs) has garnered significant interest. In such cases, weight and design cost can be decreased [1–3]. Furthermore, the control study of such special systems with fewer actuators may represent a solution in the case of an actuator failure. This ensures the effective functioning of the system and guarantees the continuity of challenging missions [4]. However, this aspect introduces a substantial problem from a control perspective, especially when compared to fully actuated systems. Underactuated systems typically exhibit nonlinearity. They often display non-minimum phase behavior, and the Lagrangian dynamics may involve non-holonomic constraints [5]. Consequently, the design of control strategies for such systems poses theoretical problems that cannot be addressed using conventional and classical control techniques. To address this issue, certain studies in the literature opt to express the system by using the Port-Controlled Hamiltonian (PCH) formulation. This formulation holds the advantage of encompassing a diverse class of physical systems [6,7] and revealed some structural properties such as symmetries, conservation laws, and stability properties. This may provide an insight into the system's behavior and facilitate the design of an effective control approach. This formulation provides the relation between the dynamics and the energy of the system, in other terms, the coupling between the non-damping and damping elements [8]. An alternative approach to Passivity-Based

Control of Underactuated Mechanical Systems. https://doi.org/10.1016/B978-0-44-324020-1.00013-0 

Control (PBC), known as Interconnection and Damping Assignment-Passivity Based Control (IDA-PBC), was introduced in [9] for the regulation of underactuated Port-Controlled Hamiltonian (PCH) systems. It has found application in the control of diverse underactuated and other systems such as the inertia wheel inverted pendulum, the ball and beam system [9], power converters [10], shunt active power filters [11], induction motors [12], FC/UC hybrid electric vehicles [13], chemical reactors [14], unmanned aerial vehicles [15], and spacecraft [16].

Numerous theoretical extensions and improvement have been proposed in the literature. The effect of physical damping on the stability and passivity of UMSs controlled by IDA-PBC was discussed in [17]. Passivation through interconnection assignment and passivation through damping injection, were proposed. These approaches underwent testing through simulation, evaluating their effectiveness on both the ball and beam system with Coulomb friction and the Vertical Take-off and Landing (VTOL) aircraft. It is noteworthy that resolving partial differential equations (PDEs) to compute the functions of kinetic and potential energy poses a challenge. The author in [18] proposed a method to simplify the partial differential equation associated with the kinetic energy via coordinate changes and re-parametrization of the target dynamics. Another parametrization of the closed-loop inertia matrix was proposed in [19]. It is an alternative simplification of the partial differential equation (PDE) related to potential energy. This approach has been implemented to address the global stabilization design of a rotary inverted pendulum. Addressing the constraints on actuator torque and aiming for a controlled action within bounds, an adjustment to the IDA-PBC controller was suggested in [20]. This modification involved incorporating a tangent function into the control law. In [21], an implementation of the discrete IDA-PBC approach was proposed for separable Hamiltonian dynamics, where the inertia matrix remains constant. This methodology was tested for the stabilization of the Inertia Wheel Inverted Pendulum (IWIP) across various sampling period values. The incorporation of Integral Control (IC) and Proportional-Integral-Derivative (PID) control has been studied in various research papers. For instance, in [22], a PID-like controller was introduced to address the robustness challenges in separable Port-Controlled Hamiltonian (PCH) systems. Assuming that the stabilization problem is resolved using the IDA-PBC method, the authors integrated an integral action to tackle robustness issues in fully actuated systems. This approach was extended to control underactuated mechanical systems, where Integral Control (IC) was applied to the non-passive output. The fundamental concept was outlined in [23], involving the addition of integral action to non-passive outputs for addressing constant and matched disturbances rejection in underactuated mechanical systems, validated through real-time experiments. Furthermore, a Model Reference Adaptive Control (MRAC) combined with a IDA-PBC controller was proposed in [6,7]. This approach demonstrated superior compensation for external disturbances compared to the standard IDA-PBC method proposed in [9]. The efficacy of this approach was validated through simulations and real-time experiments, illustrating its effectiveness in stabilizing the Inertia Wheel Inverted Pendulum under both matched and unmatched disturbances. In this chapter, we present a comparative analysis of two robust methodologies: the nonlinear Proportional-Integral (PI) controller introduced in [23] and the Model Reference Adaptive Control proposed in [7]. It is noteworthy that these controllers were effectively evaluated through experiments in comparison to the standard version of Interconnection and Damping Assignment-Passivity Based Control (IDA-PBC) proposed in [9].

## 5.2 Standard IDA-PBC controller

### 5.2.1 Control law

The equations of motion of an underactuated mechanical system can be expressed in the Port-Controlled Hamiltonian (PCH) model as:

$$\begin{bmatrix} \dot{q} \\ \dot{p} \end{bmatrix} = \begin{bmatrix} 0_{n\times n} & I_n \\ I_n & 0_{n\times n} \end{bmatrix} \nabla H(q,p) + \begin{bmatrix} 0_{n\times m} \\ G(q) \end{bmatrix} u \tag{5.1}$$

$$y = G(q)^T \nabla_p H \tag{5.2}$$

where $(q, p) \in \mathbb{R}^n$ denote, respectively, the generalized position and momenta, $u \in \mathbb{R}^m$ represent the control input vector, $G(q) : \mathbb{R}^n \to \mathbb{R}^{n\times m}$ is the input matrix with $rank(G) = m < n$, $y$ is the output vector, and $H : \mathbb{R}^n \times \mathbb{R}^n \to \mathbb{R}$ is the total energy. This last one is defined as the sum of the kinetic and potential energies, expressed as follows:

$$H(q,p) := \frac{1}{2} p^T M^{-1}(q) p + V(q) \tag{5.3}$$

where $M : \mathbb{R}^n \to \mathbb{R}^{n\times n}$ is a positive-definite inertia matrix, $M(q) = M(q)^T > 0$ and $V(q) : \mathbb{R}^n \to \mathbb{R}$ is the potential energy. The IDA-PBC control law is expressed as the sum of two terms, namely $u_{es}$ and $u_{di}$:

$$u_{ida} = u_{es}(q,p) + u_{di}(q,p) \tag{5.4}$$

where the first one is the energy shaping control. Its role is to assign the desired equilibrium $(q^*, 0)$, found by solving a set of PDEs equations, while the closed-loop system is lossless [19], for more details the reader can refer to [9]:

$$u_{es} = (G^T G)^{-1} G^T (\nabla_q H - M_d M^{-1} \nabla_q H_d + J_2 M_d^{-1} p) \tag{5.5}$$

where $M_d \in \mathbb{R}^{n\times m}$ is a positive-definite desired inertia matrix, $H_d \in \mathbb{R}^{n\times n}$ is the desired energy expressed as $H_d(q,p) = \frac{1}{2} p^T M_d(q)^{-1} p + V_d(q)$ and $V_d(q)$ is the desired potential energy verifying the condition to have an isolated minimum at $q^*$ defined by:

$$q^* = \arg\min(V_d) \tag{5.6}$$

$J_2$ is a free tuned parameter that fulfills the skew-symmetry condition, that is:

$$J_2(q,p) = -J_2^T(q,p) \tag{5.7}$$

The role of the second term of the control law (5.4) is injecting damping to achieve asymptotic stability, accomplished through negative feedback of the passive output. Its expression is described as:

$$u_{di} = (-K_v)G^T \nabla_p H_d \tag{5.8}$$

where $u_{di} \in \mathbb{R}^m$ and $K_v = K_v^T > 0$. The desired (closed-loop) PCH dynamics entail a specified equilibrium point at $(q^*, 0)$. It is expressed as:

$$\begin{pmatrix} \dot{q} \\ \dot{p} \end{pmatrix} = (J_d(q,p) - R_d(q,p))\begin{pmatrix} \nabla_q H_d \\ \nabla_p H_d \end{pmatrix} \tag{5.9}$$

where $J_d(q,p) = -J_d^T(q,p)$ denotes the interconnection matrix and $R_d = R_d^T$ represents the damping matrix. Their equations are, respectively, expressed as:

$$J_d = \begin{pmatrix} 0 & M^{-1}M_d \\ -M_d M^{-1} & J_2 \end{pmatrix} \tag{5.10}$$

and

$$R_d = \begin{pmatrix} 0 & 0 \\ 0 & GK_vG^T \end{pmatrix} \tag{5.11}$$

The closed-loop system in (5.9) has a stable equilibrium point at $(q^*, 0)$ with the Lyapunov function:

$$H_d(q,p) := \frac{1}{2}p^T M_d^{-1}(q)p + V_d(q) \tag{5.12}$$

where $V_d$ is the desired potential energy verifying the condition (5.6). The first derivative of (5.12) is expressed by:

$$\dot{H}_d(q,p) = -\|G^T M_d{}^{-1} p\|_{k_p}{}^2 \leq 0 \tag{5.13}$$

### 5.2.2 PI-IDA-PBC controller

The nonlinear PI controller consists in an outer-loop controller designed in order to solve the problem of constant disturbance rejection for underactuated mechanical systems. In accordance with [23], the control law is defined as the sum of the standard IDA-PBC controller and mathematical expression of the PI controller $\beta(q,\zeta)$:

$$u_{pi} = u_{ida} + \beta(q,\zeta) \tag{5.14}$$

The objective is to design a controller $\beta(q, \zeta)$ in order to ensure asymptotic stability of the desired equilibrium $(q^*, \zeta^*)$:

$$\beta(q, \zeta) = -K_2 K_I K_2^T G^T M^{-1} \nabla V_d - K_p K_I \zeta \tag{5.15}$$

where $K_p > 0$, $K_I > 0$ are constant matrices. The dynamics of the controller state $\zeta \in \mathbb{R}^n$ is expressed as follows:

$$\dot{\zeta} = K_2^T G^T M^{-1} \nabla V_d \tag{5.16}$$

The expression of the matrix $K_2$ includes the input matrix $G$ and the desired inertia matrix $M_d$:

$$K_2 = (G^T M_d^{-1} G)^{-1} \tag{5.17}$$

This approach requires some assumptions that are not satisfied for all kinds of underactuated mechanical systems. For more details about the proof of stability, the reader can refer to [23].

## 5.3 Model reference adaptive IDA-PBC controller

Considering that the gains values of the IDA-PBC control law $u_{ida}$ are assumed to be uncertain, the MRA-IDA-PBC controller is a direct model reference adaptive controller in which the parameters of the IDA-PBC controller are estimated online. The idea is to ensure better performances and to deal with matched disturbances. The control law is given by:

$$u_{MRA} = u_{ida} + \Delta(x, t)\hat{z} \tag{5.18}$$

where $u_{ida}$ denotes the IDA-PBC control law expressed in (5.4), $\Delta(x, t)$ is a matrix of defined functions, $z = [z_1 z_2 ... z_p]^T$ is the vector of unknown parameters and $\hat{z}$ is the estimate of $z$ obtained by the following adaptation law:

$$\dot{\hat{z}} = -K_a \Delta(x, t) y \tag{5.19}$$

where $K_a$ is chosen as a diagonal positive-definite matrix. The estimation error is then expressed as:

$$\tilde{z} = \hat{z} - z \tag{5.20}$$

The closed-loop system with the error $\tilde{z}$ can be written as follows:

$$\begin{bmatrix} \dot{q} \\ \dot{p} \\ \dot{\tilde{z}} \end{bmatrix} = \begin{bmatrix} 0 & M^{-1} M_d & 0 \\ -M_d M^{-1} & J_2 - G k_v G^T & G \Delta K_a \\ 0 & -K_a \Delta^T G^T & 0 \end{bmatrix} \begin{bmatrix} \nabla_q \bar{H} \\ \nabla_p \bar{H} \\ \nabla_z \bar{H} \end{bmatrix} \tag{5.21}$$

where the Hamiltonian $\bar{H}$ is given by:

$$\bar{H}(q,p,\tilde{z}) = H_d(q,p) + \frac{1}{2}\tilde{z}^T K_a^{-1}\tilde{z} \tag{5.22}$$

The dynamic model of IWIP can be obtained using the Euler–Lagrange method [24–29]. The equations of motion can be expressed by:

$$\begin{bmatrix} a+I_{WC} & I_{WC} \\ I_{WC} & I_{WC} \end{bmatrix}\begin{bmatrix} \ddot{\theta}_1 \\ \ddot{\theta}_2 \end{bmatrix} - \begin{bmatrix} bg\sin(\theta_1) \\ 0 \end{bmatrix} = \begin{bmatrix} 0 \\ u \end{bmatrix} \tag{5.23}$$

where $\theta = [\theta_1, \theta_2]^T$, is the vector of generalized positions, $u$ is the torque generated by the actuator and acting on the inertia wheel. In the above model, the terms related to the friction in the pendulum joint are neglected. $a = ML^2 + I_{PB}$ and $b = ml + ML$. The dynamic parameters of the system $I_{PC}$ and $I_{PB}$ denote the rotational inertia of the pendulum about the pendulum center of mass (PC) and the rotational inertia of the pendulum about its base (PB), respectively. $I_{WC}$ represents the rotational inertia of the wheel about its center of mass (WC). $l$ denotes the length from the pendulum base (PB) to the pendulum center of mass (PC), the length from the pendulum base (PB) to the wheel center of mass (WC) is denoted by $L$, $m$ and $M$ are the mass of the pendulum and wheel, respectively, and $g$ is the constant of gravitational acceleration.

The first step is to transform the system model in PCH form. Consider the following change of coordinates [30]:

$$q_1 = \theta_1 \tag{5.24}$$
$$q_2 = \theta_1 + \theta_2 \tag{5.25}$$
$$p_1 = a\dot{q}_1 \tag{5.26}$$
$$p_2 = I_{WC}\dot{q}_2 \tag{5.27}$$

We obtain the following simplified model with a diagonal inertia matrix:

$$\begin{pmatrix} a & 0 \\ 0 & I_{WC} \end{pmatrix}\begin{pmatrix} \ddot{q}_1 \\ \ddot{q}_2 \end{pmatrix} - \begin{pmatrix} bg\sin(q_1) \\ 0 \end{pmatrix} = \begin{pmatrix} -1 \\ 1 \end{pmatrix} u \tag{5.28}$$

The obtained model is rewritten through Hamilton's equations of motion as:

$$\begin{pmatrix} \dot{q}_1 \\ \dot{q}_2 \\ \dot{p}_1 \\ \dot{p}_2 \end{pmatrix} = \begin{pmatrix} \frac{p_1}{a} \\ \frac{p_2}{I_{WC}} \\ bg\sin(q_1) - u \\ u \end{pmatrix} \tag{5.29}$$

where $q = [q_1, q_2]^T$ and $p = [p_1, p_2]^T$ are the generalized positions and momenta, respectively.

The IDA-PBC control law can be described as [9]:

$$u_{ida} = \gamma_1 \sin q_1 + k_p(q_2 + \gamma_2 q_1) + k_v k_2(\dot{q}_2 + \gamma_2 \dot{q}_1) \tag{5.30}$$

$$u_{ida} = \gamma_1 \sin q_1 + K_1 q_1 + K_2 q_2 + K_3 p_1 + K_4 p_2 \tag{5.31}$$

where $k_p > 0$ is a proportional gain, $k_v > 0$ is a damping injection gain, $\gamma_1 > b$, $\gamma_2 = \frac{a\gamma_1}{I_{WC}(\gamma_1 - b)}$ and $k_2 > 0$:

$$K_1 = k_p \gamma_2 \tag{5.32}$$

$$K_2 = k_2 \tag{5.33}$$

$$K_3 = \frac{k_v k_2 \gamma_2}{a} \tag{5.34}$$

$$K_4 = \frac{k_v k_2}{I_{WC}} \tag{5.35}$$

The MRA-IDA-PBC control law can be described as [7,31]:

$$u_{MRA} = u_{ida} + \Delta(x,t)\hat{z} \tag{5.36}$$

where

$$\Delta(x,t) = [p_2, p_1]^T \tag{5.37}$$

$$\hat{z} = [\hat{K}_3, \hat{K}_4] \tag{5.38}$$

The adaptation law is expressed as follows:

$$\dot{\hat{z}} = -K_a \Delta(x,t)^T G(x)^T \Delta_p H_d$$

where $K_a$ is a diagonal positive-definite matrix.

The proposed approach is a revision of the adaptive version of the MRA-IDA-PBC controller. In the previous approach only two parameters are estimated online by the adaptation law. In our work, we propose to estimate three parameters instead of two. We propose to change the model reference adaptive structure of $\Delta(x,t)$ and $\hat{z}$ as:

$$\Delta(x,t) = [p_2, p_1, q_1]^T \tag{5.39}$$

$$\hat{z} = [\hat{K}_{2e}, \hat{K}_{3e}, \hat{K}_{4e}] \tag{5.40}$$

## 5.4 Closed-loop stability analysis

The stability analysis is demonstrated by choosing $\bar{H}(q,p,\tilde{z})$ as a Lyapunov candidate function as demonstrated in [7]:

$$\dot{\bar{H}} = \dot{H}_d + K_a^{-1}\tilde{z}\dot{\tilde{z}} \tag{5.41}$$

$$\dot{\bar{H}} = \dot{H}_d + K_a^{-1}\bar{z}\Delta^T(x,t)G^T\nabla_p H_d \tag{5.42}$$

$$\dot{\bar{H}} = \nabla_q H_d\dot{q} + \nabla_p H_d\dot{p} + K_a^{-1}\bar{z}\Delta^T(x,t)G^T\nabla_p H_d \tag{5.43}$$

$$\dot{\bar{H}} \leq -\alpha|\nabla_p H_d^T G|^2 \tag{5.44}$$

where $\alpha$ is a positive constant.

## 5.5 Numerical simulation results

In this section, the numerical simulation results obtained by the proposed MRA-IDA-PBC control solution are demonstrated and discussed.

### 5.5.1 Scenario 1: Punctual disturbance rejection

In this section, we will test and verify the robustness of the proposed controller compared with the standard IDA-PBC controller [9] and the MRA-IDA-PBC proposed in [7]. The control parameters were selected as $\gamma_1 = 6.1284$, $k_1 = 1.0367$, $k_2 = 0.0011$, $k_3 = 16.62$, $k_4 = 3.4640$, $K_3(0) = 2$, $K_4(0) = 10$, $K_{3e}(0) = 1$, $K_{4e}(0) = 10$, $K_{2e}(0) = 15.5$, $K_a = \begin{pmatrix} 2.08e-4 & 0 \\ 0 & 2.08e-4 \end{pmatrix}$, $K_{a2} = \begin{pmatrix} 48e-1 & 0 & 0 \\ 0 & 1e-4 & 0 \\ 0 & 0 & 0.001 \end{pmatrix}$. Two different simulation scenarios are proposed. In the first one, the system is subject to two punctual disturbances introduced at time instants $t = 10$ s and $t = 30$ s. In the second one, the system is subject to a persistent perturbation applied at time instant $t = 10$ s. The first input disturbance is a constant torque equal to 0.8 N m and the second is equal to 2.5 N m. The evolution of the system states is depicted in Fig. 5.1. Fig. 5.2 displays the evolution of estimated parameters versus time. It can be observed that the first MRA-IDA-PBC controller is better than the standard IDA-PBC controller in terms of compensating matched disturbances, as discussed in [7]. However, the proposed MRA-IDA-PBC ensures a faster convergence towards the equilibrium point and a better performance toward punctual disturbances. Moreover, the pendulum position remains in the interval $[-0.1, 0.05]$ rad, in contrast to the standard IDA-PBC and the first MRA-IDA-PBC controller. The disadvantage of the proposed controller can be noted in the evolution of the control input, see Fig. 5.3. Accordingly, the proposed controller produces a high control input to ensure the convergence to the unstable equilibrium point. This may result in a high energy consumption, exceeding the limits of the real actuator.

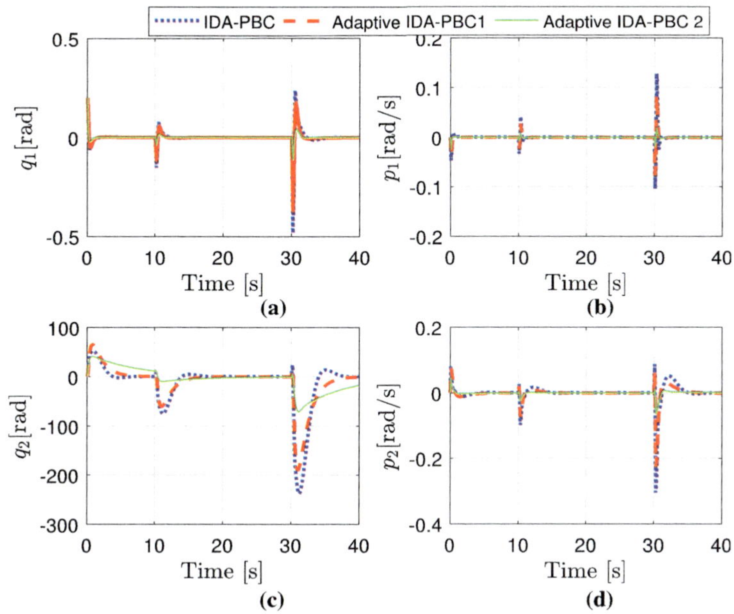


**FIGURE 5.1** Comparative results of advanced IDA-PBC-based controllers for scenario 1: Punctual disturbance rejection. **(a)** Pendulum position, **(b)** Angular velocity of the pendulum, **(c)** Inertia wheel position, **(d)** Angular velocity of the inertia wheel.

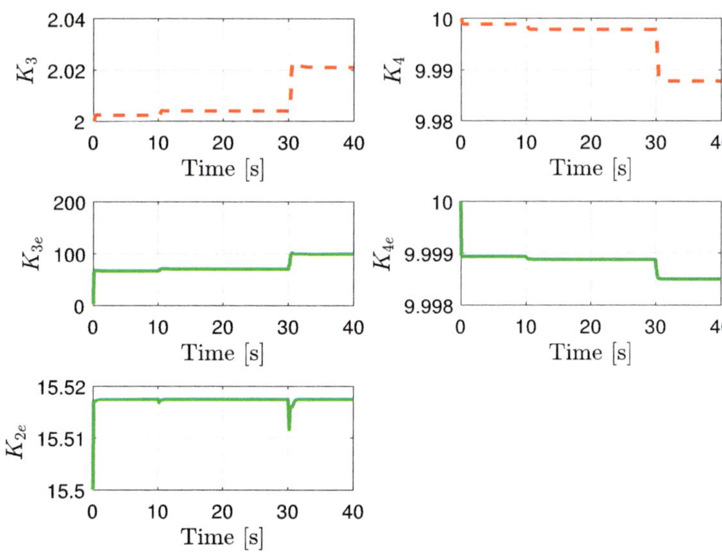


**FIGURE 5.2** Evolution of the estimated parameters of adaptive IDA-PBC-based controllers for scenario 1: Punctual disturbance rejection.

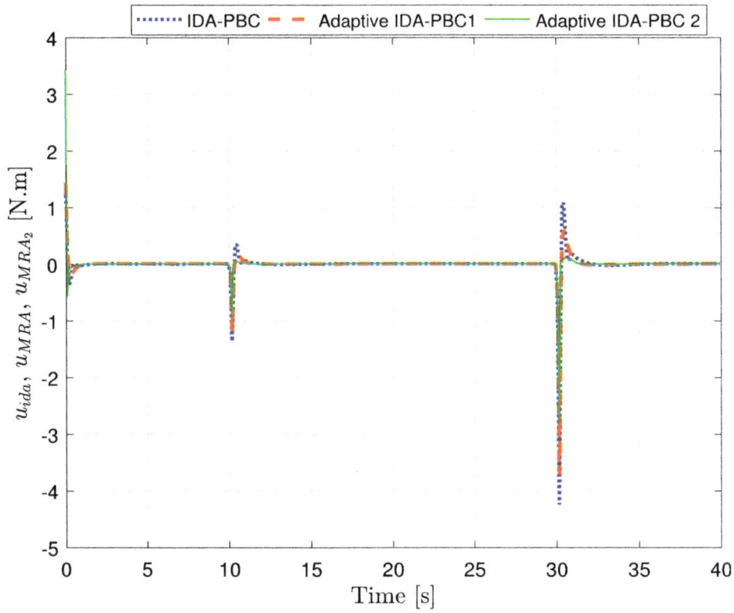


**FIGURE 5.3** Evolution of the control input of advanced IDA-PBC-based controllers for scenario 1: Punctual disturbance rejection.

### 5.5.2 Scenario 2: Persistent disturbance rejection

In this scenario, we consider the problem of persistent disturbance rejection in the case of the IWIP. The applied disturbance is a constant perturbation, equal to 0.7 N m, applied during the time interval [5 s, 40 s]. The resulted system states for the three controllers are displayed in Fig. 5.4. It can be noted that the states of the system converge to zero for all controllers. However, better performance in terms of disturbance rejection is displayed for the case of the proposed revisited adaptive controller. The effect of persistent disturbance may be observed in the evolution of the estimated parameters, depicted in Fig. 5.5.

## 5.6 Real-time experimental results

In this section, we will conduct experimental comparisons to evaluate the performance of controllers based on Interconnection and Damping Assignment-Passivity Based Control (IDA-PBC). Two experimental scenarios were conducted on the described testbed. In the first scenario, no external disturbances were taken into account. In the second scenario, an extra mass was affixed to the pendulum body, thereby introducing a continuous external disturbance affecting the passive joint of the pendulum.

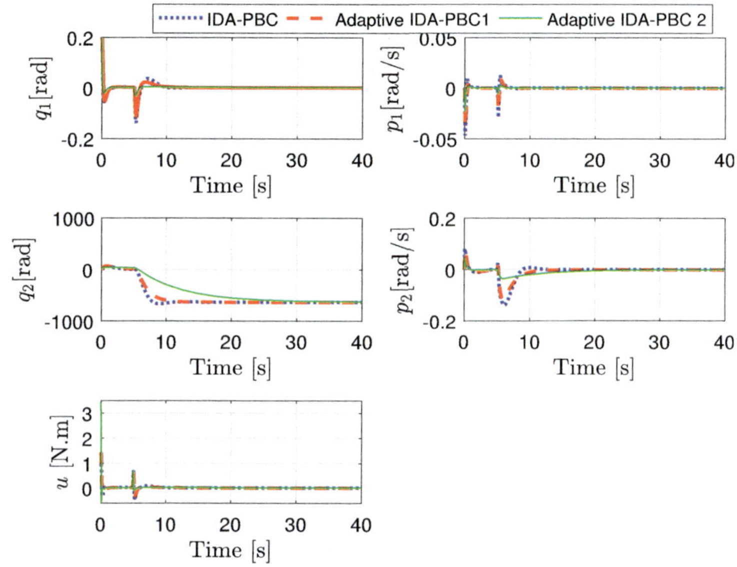


**FIGURE 5.4** Comparative results of advanced IDA-PBC-based controllers for scenario 2: Persistent disturbance rejection.

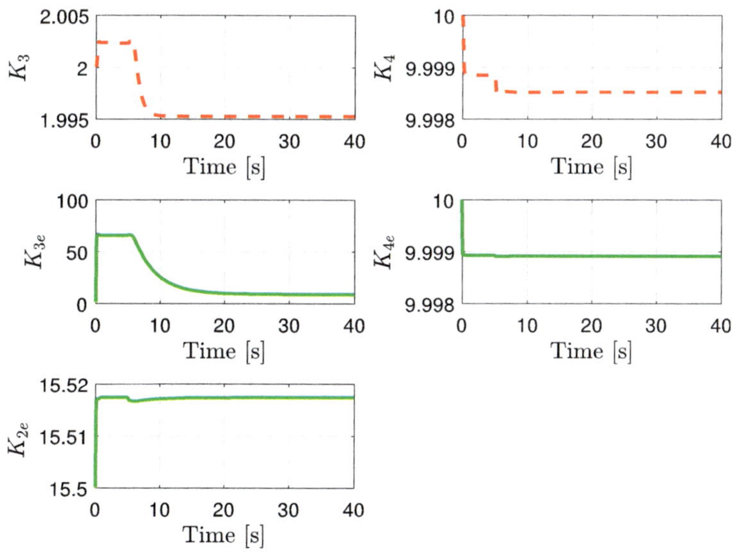


**FIGURE 5.5** Evolution of the estimated parameters of adaptive IDA-PBC-based controllers for scenario 2: Persistent disturbance rejection.

### 5.6.1 Scenario 1: Nominal case

The control design parameters and initial conditions are summarized in Tables 5.1 and 5.2. The evolution of angular positions and velocities for the

**TABLE 5.1 Summary of control design parameters.**

| Parameter | Value | Parameter | Value |
|---|---|---|---|
| $a$ | 0.032 | $K_a$ | [0.025,0; 0,0.025] |
| $I_{WC}$ | 0.000417 | $\hat{K}_3(0)$ | 3 |
| $\gamma_1$ | 6.120 | $\hat{K}_4(0)$ | 8 |
| $K_1$ | 0.0471 | $\zeta(0)$ | 0.18 |
| $K_2$ | 0.0005 | $K_I$ | 0.18 |
| $K_3$ | 16.4 | $K_P$ | 3.42 |
| $K_4$ | 3.4 | $K_2$ | 1.07 |
| $\alpha$ | 1.8934 | | |

**TABLE 5.2 Experimental configurations for MRA-IDA-PBC and PI-IDA-PBC controllers.**

| Controller | Initial conditions |
|---|---|
| IDA-PBC | $(q_1, q_2, p_1, p_2)^T = (0.17, 0, 0, 0)^T$ |
| MRA-IDA-PBC | $(q_1, q_2, p_1, p_2, \hat{K}_3, \hat{K}_4)^T = (0.17, 0, 0, 0, \hat{K}_3(0), \hat{K}_4(0))^T$ |
| PI-IDA-PBC | $(q_1, q_2, p_1, p_2, \zeta)^T = (0.17, 0, 0, 0, \zeta(0))^T$ |

standard IDA-PBC, the model reference IDA-PBC, and the PI-IDA-PBC are illustrated in Fig. 5.6. The phase portrait $(\theta_1, \dot{\theta}_1)$ is depicted in Fig. 5.7. The control input and the estimated gain are illustrated in Figs. 5.8 and 5.9, respectively. The superior convergence of the PI-IDA-PBC controller compared to the MRA-IDA-PBC and IDA-PBC controllers is noted. The evolution of the control input with PI-IDA-PBC converges more rapidly to a steady state value and exhibits fewer oscillations than the control input of the MRA-IDA-PBC and IDA-PBC controllers. The figures illustrate the convergence of the estimated gain of the adaptive controller and $\zeta$ parameters to their steady state values. The phase portrait provides a visual representation of the trajectories of the dynamic system. Notably, starting from the same initial condition, the PI-IDA-PBC controller follows the shortest path.

In order to measure the effectiveness of the best controllers, we opt for the evaluation criteria based on metrics such the Root Mean Squared Error, Integral of Squared Error, Integral of Absolute Error, Integral of Time-weighted Absolute Error, and Input-Torque Criteria, see Table 5.3. The RMSE measures the average magnitude of the differences between predicted and actual angular positions. A lower RMSE indicates better accuracy. In this case, PI-IDA-PBC has a lower RMSE, suggesting it provides more accurate predictions in terms of angular positions. ISE quantifies the total squared error over time. A lower ISE indicates better performance. Here, PI-IDA-PBC has a slightly lower ISE, suggesting it performs better in minimizing the squared error. IAE measures the total absolute error over time. Similar to ISE, a lower IAE is better. In this case,

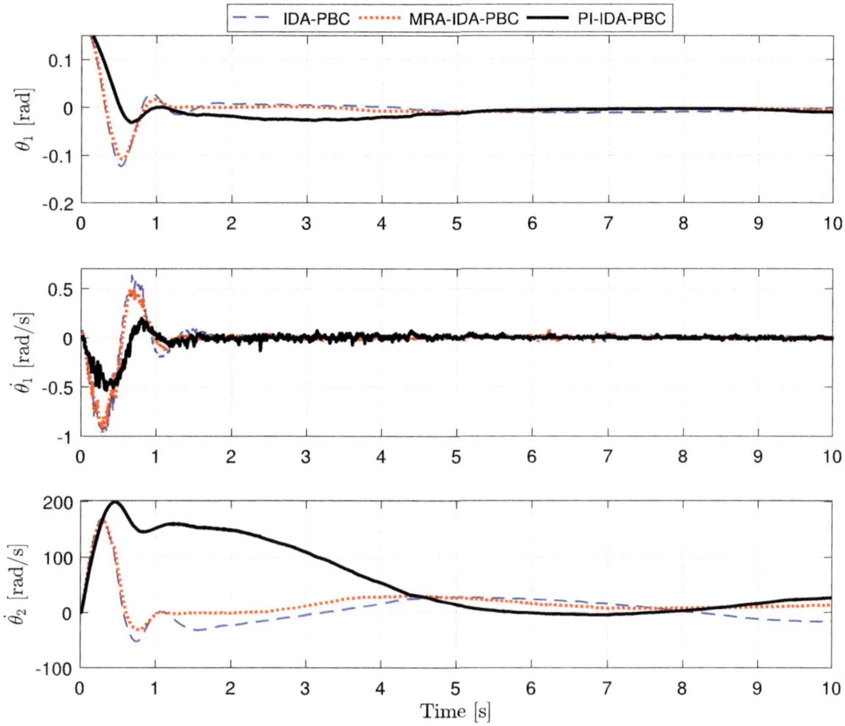


**FIGURE 5.6** Obtained experimental results for scenario 1.

**TABLE 5.3** Quantification of the performance through different evaluation criteria for scenario 1.

| Criteria | Scenario 1 | |
|---|---|---|
| | MRA-IDA-PBC | PI-IDA-PBC |
| RMSE [rad] | 0.0035 | 0.0012 |
| ISE | 0.0098 | 0.0076 |
| IAE | 0.1951 | 0.1885 |
| ITAE | 0.0142 | 0.0049 |
| $E_\tau$ [N m] | 124.6523 | 89.2769 |

PI-IDA-PBC has a slightly lower IAE, indicating better performance in minimizing the absolute error. A lower ITAE suggests better overall performance. Here, PI-IDA-PBC significantly outperforms MRA-IDA-PBC in terms of time-weighted absolute error. PI-IDA-PBC requires less control effort, suggesting it achieves better performance with less energy consumption.

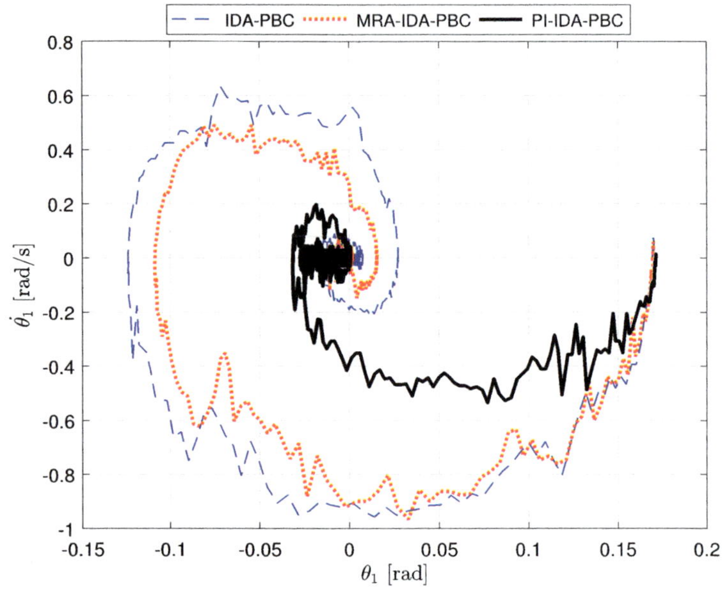


**FIGURE 5.7** Phase portrait $(\theta_1, \dot{\theta}_1)$ for scenario 1.

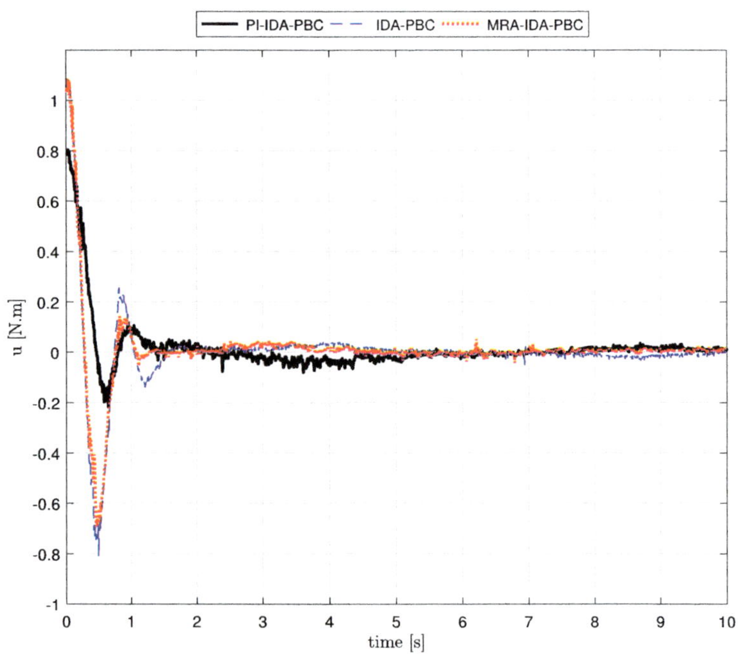


**FIGURE 5.8** The evolution of the control input for scenario 1.

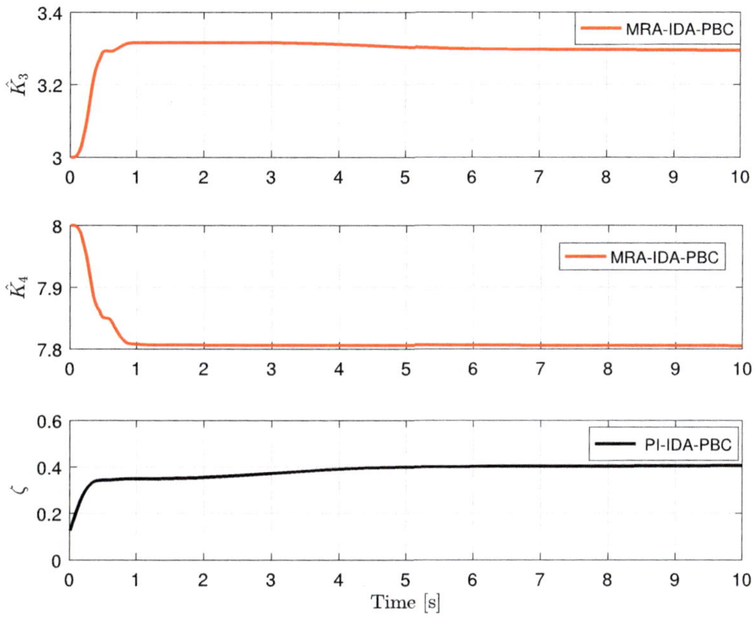


**FIGURE 5.9** Evolution of the estimated gains for scenario 1.

To sum up, the PI-IDA-PBC controller outperforms the MRA-IDA-PBC controller across all the evaluated criteria in Scenario 1. It demonstrates better accuracy, error minimization, and control efficiency.

### 5.6.2 Scenario 2: Persistent disturbances rejection

The disturbance considered in this scenario is persistent and continuously applied to the inverted pendulum in the form of an additional mass attached to the pendulum body before the experiment. This can be represented as an external force $F_{\text{dis}}$ acting on the inverted pendulum, generating a torque $\tau_{\text{ext}}$ around the pivot of the passive joint. In this section, we investigate the capability of the implemented controllers to reject this external disturbance. The evolution of the angular position and velocity of the inertia wheel obtained for the second scenario is presented in Fig. 5.10. Figs. 5.11 and 5.12 illustrate the evolution of the control input and estimated gains of the controllers. Effectively, the IDA-based controllers reject the persistent disturbance introduced and maintain the system in proximity to its unstable equilibrium point. With the PI controller, the actuator rotation is continuous, and the disturbance has a minimal impact on the angular velocity evolution of the inertia wheel. Nevertheless, with the adaptive and standard controllers, the influence of the persistent disturbance is noted in the behavior of the angular velocity of the inertia wheel, noticeable as a deviation in the angular velocity trajectory.

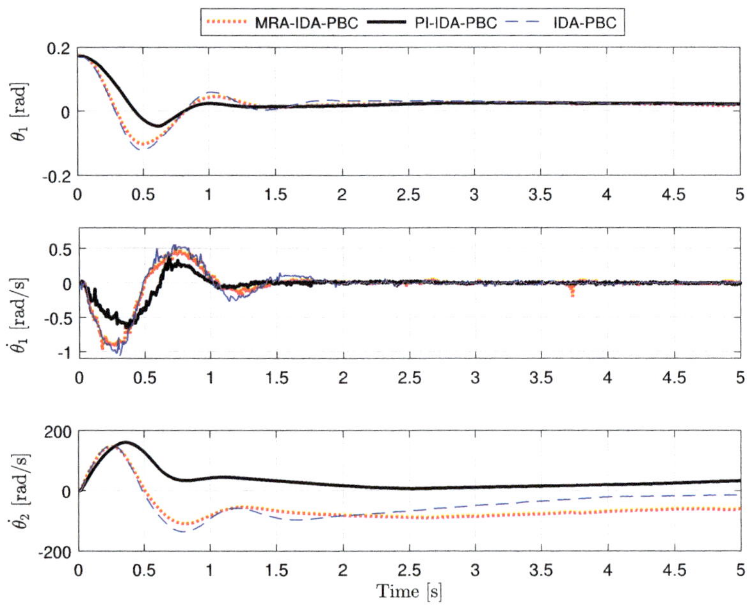


**FIGURE 5.10** Obtained experimental results for scenario 2.

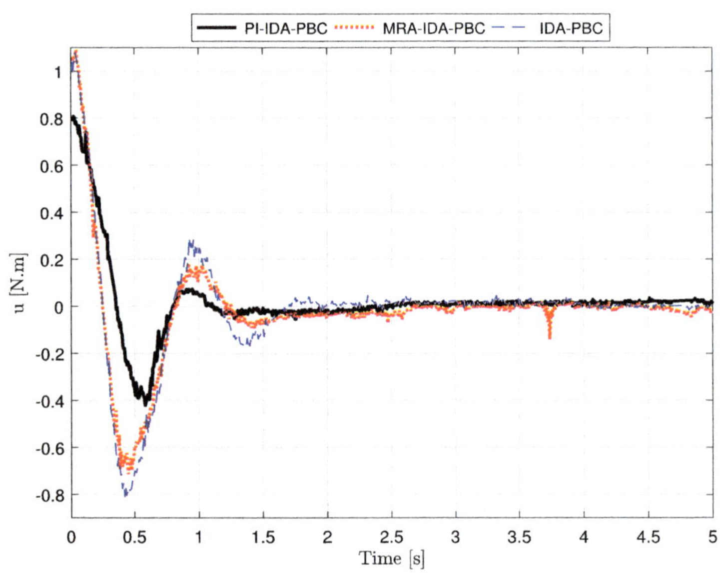


**FIGURE 5.11** The evolution of the control input for scenario 2.

The experimental results and the comparative study using various performance criteria presented in Table 5.4 demonstrate that the PI-IDA-PBC con-

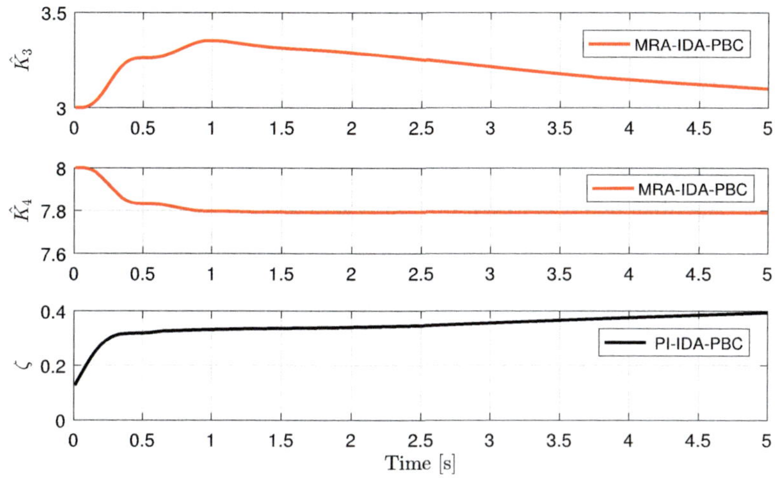


**FIGURE 5.12** Evolution of the estimated gains for scenario 2.

**TABLE 5.4 Quantification of the performance through different evaluation criteria for scenario 2.**

| Criteria | Scenario 2 | |
|---|---|---|
| | MRA-IDA-PBC | PI-IDA-PBC |
| RMSE [rad] | 0.0199 | 0.0187 |
| ISE | 0.0198 | 0.0171 |
| IAE | 0.6656 | 0.5822 |
| ITAE | 0.2228 | 0.1655 |
| $E_\tau$ [N m] | 126.199 | 112.202 |

troller ensures superior convergence compared to the MRA-IDA-PBC controller.

The implemented controllers consistently apply a DC voltage to the motor, ensuring its continuous rotation to counteract the disturbing torque and, consequently, enabling the rejection of the external disturbance. A lower control input is obtained for the case of the PI controller compared to the adaptive and standard one. This result is confirmed when computing the input torque metric.

## 5.7 Conclusion

This chapter presents an overview of passivity-based control named the IDA-PBC controller, with an experimental comparative study of three controllers, namely the IDA-PBC, MRA-IDA-PBC, and PI-IDA-PBC controllers. The first one is the standard controller, the second is an adaptive control scheme combined with the IDA-PBC design methodology, while the latter is an outer-loop controller added to the interconnection and damping assignment passivity-based control.

Both adaptive and PI-based controllers aim to improve the robustness of the standard IDA-PBC for underactuated mechanical systems. A comparative study was conducted experimentally using the inertia wheel inverted pendulum (IWIP). The objective was to stabilize the IWIP around its unstable equilibrium point and maintain it in this state despite the presence of external disturbances. Two experimental scenarios were considered: control of the nominal system without perturbations and control under an external persistent perturbation.

Based on the obtained experimental results and various performance-evaluation criteria, the PI-IDA-PBC demonstrated superior performance in terms of convergence and energy consumption compared to the MRA-IDA-PBC controller. Future work may explore the combination of the adaptation law with the PI controller to estimate the proportional and integral gains dynamically. Additionally, discussions can be initiated regarding the generalization of this study to other classes of underactuated mechanical systems. Furthermore, an investigation into automatic optimal tuning of control design parameters for both controllers could be explored.

## References

[1] S. Andary, A. Chemori, S. Krut, Control of the underactuated inertia wheel inverted pendulum for stable limit cycle generation, Advanced Robotics 23 (2009) 1999–2014.

[2] A. Hfaiedh, A. Chemori, A. Abdelkrim, A revisited adaptive super-twisting control of underactuated mechanical systems: design and experiments, Transactions of the Institute of Measurement and Control (2024), https://doi.org/10.1177/01423312241274060.

[3] A. Hfaiedh, A. Chemori, A. Abdelkrim, Observer-based robust integral of the sign of the error control of class I of underactuated mechanical systems: theory and real-time experiments, Transactions of the Institute of Measurement and Control 44 (2) (2022) 339–352, https://doi.org/10.1177/01423312211031396.

[4] A. Choukchou-Braham, B. Cherki, M. Djemai, K. Busawon, Analysis and Control of Underactuated Mechanical Systems, Springer International Publishing, Switzerland, 2014.

[5] S. Krafes, Z. Chalh, A. Saka, A review on the control of second order underactuated mechanical systems, Complexity (2018) 17.

[6] N. Khraief Haddad, A. Chemori, J.J. Pena, S. Belghith, Stabilization of inertia wheel inverted pendulum by model reference adaptive IDA-PBC: from simulation to real-time experiments, in: 2015 3rd International Conference on Control, Engineering Information Technology (CEIT), 2015, pp. 1–6.

[7] N. Khraief Haddad, A. Chemori, S. Belghith, Robustness enhancement of IDA-PBC controller in stabilising the inertia wheel inverted pendulum: theory and real-time experiments, International Journal of Control 91 (12) (2018) 2657–2672.

[8] M. Ryalat, D. Laila, A robust IDA-PBC approach for handling uncertainties in underactuated mechanical systems, IEEE Transactions on Automatic Control 63 (2018) 3495–3502.

[9] R. Ortega, M. Spong, F. Gomez-Estern, G. Blankenstein, Stabilization of a class of underactuated mechanical systems via interconnection and damping assignment, IEEE Transactions on Automatic Control 47 (2002) 1218–1233.

[10] H. Rodriguez, R. Ortega, G. Escobar, N. Barabanov, A robustly stable output feedback saturated controller for the boost DC-to-DC converter, Systems & Control Letters 40 (2000) 1–8.

[11] F. Serra, C. De Angelo, Direct power control of a shunt active power filter using a modified IDA-PBC approach with integral action, IEEE Transactions on Circuits and Systems. II, Express Briefs 70 (2023) 1991–1995.
[12] H. Gonzalez, M. Duarte-Mermoud, I. Pelissier, J. Travieso, R. Ortega, A novel induction motor control scheme using IDA-PBC, Stability and Control: Theory and Applications 6 (2008) 59–68.
[13] M.R. Naseh, A. Behdani, Non-linear control of fuel cell/ultra-capacitor hybrid electric vehicle using comprehensive function algorithm based on IDA-PBC, Advanced Energy Conversion Materials (2023), https://doi.org/10.37256/aecm.4120231947.
[14] H. Ramírez, D. Sbarbaro, R. Ortega, On the control of non-linear processes: an IDA-PBC approach, Journal of Process Control 19 (2009) 405–414.
[15] M.E. Guerrero-Sánchez, O. Hernández-González, G. Valencia-Palomo, et al., Robust IDA-PBC for under-actuated systems with inertia matrix dependent of the unactuated coordinates: application to a UAV carrying a load, Nonlinear Dynamics 105 (2021) 3225–3238, https://doi.org/10.1007/s11071-021-06776-7.
[16] S. Aoues, D. Matignon, D. Alazard, Control of a flexible spacecraft using discrete IDA-PBC design, IFAC-PapersOnLine 48 (13) (2015) 188–193.
[17] F. Gomez-Estern, A.J. Van der Schaft, Physical damping in IDA-PBC controlled underactuated mechanical systems, European Journal of Control 10 (5) (2004) 451–468.
[18] G. Viola, R. Ortega, R. Banavar, J.A. Acosta, A. Astolfi, Total energy shaping control of mechanical systems: simplifying the matching equations via coordinate changes, IEEE Transactions on Automatic Control 52 (6) (2007) 1093–1099.
[19] M. Ryalat, D.S. Laila, IDA-PBC for a class of underactuated mechanical systems with application to a rotary inverted pendulum, in: 52nd IEEE Conference on Decision and Control, 2013, pp. 5240–5245.
[20] V. Santibanez, R. Kelly, J. Sandoval, Control of the inertia wheel pendulum by bounded torques, in: Proceedings of the 44th IEEE Conference on Decision and Control, 2005, pp. 8266–8270.
[21] F. Tiefensee, S. Monaco, D. Normand-Cyrot, IDA-PBC under sampling for port-controlled Hamiltonian systems, in: Proceedings of the 2010 American Control Conference, 2010, pp. 1811–1816.
[22] M. Ryalat, D.S. Laila, M.M. Torbati, Integral IDA-PBC and PID-like control for port-controlled Hamiltonian systems, in: 2015 American Control Conference (ACC), Chicago, United States, 2015, pp. 5365–5370.
[23] A. Donaire, J.G. Romero, R. Ortega, B. Siciliano, M. Crespo, Robust IDA-PBC for underactuated mechanical systems subject to matched disturbances, International Journal of Robust and Nonlinear Control 27 (6) (2017) 1000–1016.
[24] A. Hfaiedh, A. Chemori, A. Abdelkrim, Disturbance observer-based super-twisting control for the inertia wheel inverted pendulum, in: 17th International Multi-Conference on Systems, Signals and Devices (SSD'20), 2020, pp. 751–756.
[25] A. Hfaiedh, A. Chemori, A. Abdelkrim, RISE controller for class I underactuated mechanical systems: design and real-time experiments, in: The 3rd International Conference on Electromechanical Engineering (ICEE'2018), 2018.
[26] S. Andary, A. Chemori, M. Benoit, J. Sallantin, A dual model-free control of underactuated mechanical systems, application to the inertia wheel inverted pendulum, in: 2012 American Control Conference (ACC), 2012, pp. 1029–1034.
[27] S. Andary, A. Chemori, S. Krut, Estimation-based disturbance rejection in control for limit cycle generation on inertia wheel inverted pendulum testbed, in: 2009 IEEE/RSJ International Conference on Intelligent Robots and Systems, 2009, pp. 1302–1307.
[28] S. Andary, A. Chemori, A dual model-free control of non-minimum phase systems for generation of stable limit cycles, in: 2011 50th IEEE Conference on Decision and Control and European Control Conference, 2011, pp. 1387–1392.

[29] S. Andary, A. Chemori, S. Krut, Stable limit cycle generation for underactuated mechanical systems, application: inertia wheel inverted pendulum, in: 2008 IEEE/RSJ International Conference on Intelligent Robots and Systems, 2008, pp. 526–531.
[30] A. Hfaiedh, A. Chemori, A. Abdelkrim, Stabilization of the inertia wheel inverted pendulum by advanced IDA-PBC based controllers: comparative study and real-time experiments, in: 2020 17th International Multi-Conference on Systems, Signals & Devices (SSD), Monastir, Tunisia, 2020, pp. 753–760.
[31] N. Haddad, S. Belghith, H. Gritli, A. Chemori, From Hopf bifurcation to limit cycles control in underactuated mechanical systems, International Journal of Bifurcation and Chaos 27 (07) (2017).

Part III

# Control solutions for stable limit cycle generation problem

Chapter 6

# Partial feedback linearization and optimization

Contents



## 6.1 Introduction

The first proposed control approach[1] aims to generate stable limit cycles on all the coordinates of an underactuated mechanical system. The difficulty lies in the fact that only some of the coordinates are actuated. Indeed, conventional techniques for controlling underactuated mechanical systems, such as feedback

[1] This approach has been developed during the Ph.D thesis of Sébastien Andary to whom we are grateful for permitting us to reproduce some of his results.

exact linearization, are therefore impossible to implement. Therefore several control schemes have been proposed in the literature to deal with control of underactuated mechanical systems [1–5]. This chapter presents a control approach that has led to some publications in international conferences and scientific journals [6–8]. This control approach consists of generating periodic reference trajectories tracked on the non-actuated part of the system's coordinates. Reference trajectories are tracked using the principle of partial feedback linearization [9–11]. Under the convergence assumption, these coordinates describe a stable limit cycle [21–26] if the reference trajectories are periodic.

The dynamics of the sub-system made up of the remaining (actuated) coordinates, unstable in the case of a non-minimum-phase system [12–15], is stabilized by optimizing the reference trajectories followed on the linearized sub-system. In addition, the ability to reject disturbances is improved by estimating those disturbances and compensating for them in the proposed control law. Internal dynamics are therefore stabilized. As a result of the periodicity of the reference trajectories, the actuated coordinates sub-system is stabilized around a stable limit cycle [21–26].

## 6.2 Motivation and proposed control scheme

The proposed control approach is divided into several blocks. A part of the sub-actuated mechanical system is linearized using the principle of partial feedback linearization [9–11]. This part consists of the system's non-actuated coordinates. A family of periodic reference trajectories is defined for the non-actuated part. These reference trajectories are parameterized by a parameter used as an additional control input. This control input is used to stabilize the system's internal dynamics. At the start of each period (cycle), the value of the parameter that minimizes the divergence of the internal dynamics is chosen. Once the parameter has been chosen, the reference trajectories are tracked for one complete cycle.

The whole block diagram of this control approach is illustrated in Fig. 6.1.

The proposed control approach is presented in the general case of underactuated mechanical systems. Consider an underactuated mechanical system with $n$ degrees of freedom and $m$ actuators ($m < n$), and generalized coordinates $q^1, \ldots, q^n$. The vector of coordinates $q \in \mathbb{R}^n$ is partitioned into two parts: one consisting of the actuated coordinates $q_a \in \mathbb{R}^m$ and the other of the non-actuated coordinates $q_{na} \in \mathbb{R}^{n-m}$. Using the Euler–Lagrange method, it is possible to obtain the equations governing the dynamics of this system [16,17,27]:

$$M_{11}\ddot{q}_a + M_{12}\ddot{q}_{na} + h_1 + g_1 = u \tag{6.1}$$

$$M_{12}\ddot{q}_a + M_{22}\ddot{q}_{na} + h_2 + g_2 = 0 \tag{6.2}$$

where

$$M(q) = \begin{bmatrix} M_{11} & M_{12} \\ M_{12} & M_{22} \end{bmatrix} \tag{6.3}$$

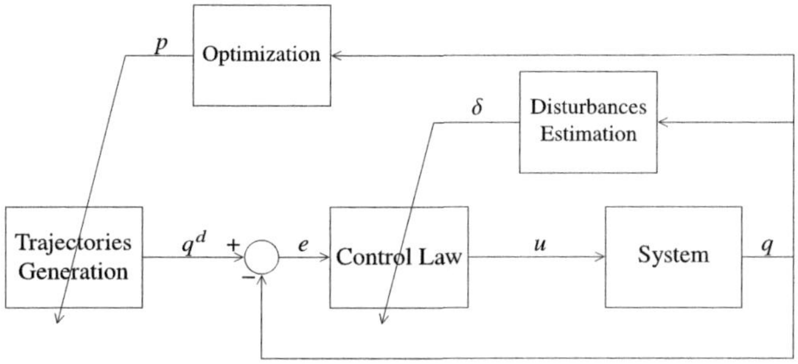


**FIGURE 6.1** General overview of the proposed control approach: Reference trajectories ($q^d$) parameterized by a certain parameter ($p$) are generated to be tracked on a part of the system. The tracking is performed by a control law ($u$) that linearizes the coordinates of this part of the system. An optimization criterion for the remaining part of the coordinates is evaluated at the end of the cycle, and the reference trajectory for the next cycle is chosen by optimizing this criterion. Perturbations are estimated and compensated in the designed control law.

is the symmetrical, positive-definite inertia matrix of the system, $h_1(q,\dot{q}) \in \mathbb{R}^m$ and $h_2(q,\dot{q}) \in \mathbb{R}^{n-m}$ are the vectors of the terms corresponding to the centrifugal and Coriolis forces, $g_1(q) \in \mathbb{R}^m$ and $g_2(q) \in \mathbb{R}^{n-m}$ are the vectors of the terms corresponding to the force of gravity; finally, $u$ designates the vector of generalized forces produced by the system's $m$ actuators (control inputs).

### 6.2.1 Partial feedback linearization principle

State feedback linearization involves linearizing the dynamics of a system [27]. A state feedback is applied to the designed system to compensate for nonlinearities in its dynamics. The closed-loop system is then equivalent to a linear system of the same dimension (same number of coordinates). In the general case, exact linearization is only possible if the system is fully actuated.

In the case of an underactuated system, it is possible to linearize only a part of the system (a subset of its coordinates). This is known as partial feedback linearization [9–11,18]. The number of coordinates of a nonlinear system that can be linearized is equal to the number of independent actuators in the system. The actuated coordinates of the system can always be linearized. The non-actuated coordinates, on the other hand, can be linearized if the system satisfies certain properties. Indeed, the system dynamics should present a strong coupling between passive (non-actuated) and active (actuated) coordinates. This property is analyzed and highlighted in [17–19], we recall it here:

**Definition 1.** The system (6.1) and (6.2) is said to be *strongly inertially coupled* if and only if $\text{Rank}(M_{12}(q)) = n - m$.

The property of strong inertial coupling implies $m \geq n - m$. In other words, the number of actuated coordinates must be greater than or equal to the number of non-actuated coordinates. The resulting closed-loop system then can be split into two sub-systems: (i) a sub-system with linear dynamics, and (ii) a sub-system with nonlinear dynamics, which is unstable if the system is non-minimum-phase [12–15]. The dynamics of the nonlinearized sub-system is called the internal dynamics of the nonlinear system. In the proposed control approach, it is the non-actuated part that is linearized.

### 6.2.2 Reference trajectory generation

Once the non-actuated part of the system has been linearized, it is possible to impose on it reference trajectories [28]. In order to obtain a stable limit cycle [21–26] on the linearized part, the reference trajectories should be periodic. We then define a set of parameterized reference trajectories, all compatible with each other. Any two trajectories in the set should be able to follow each other over two successive periods, in order to obtain a smooth overall trajectory with no discontinuities. As a result, every trajectory in the set has the same boundary conditions, i.e., the same initial and final conditions. Also, each trajectory in the set is a function of a parameter $p$ affecting a stability criterion of the internal dynamics. The range of admissible values of the optimization parameter $p$ is called $\mathcal{P}$. A reference trajectory $T$ for the non-actuated part (i.e., the linearized part) is defined for a given period $t_c$:

$$T : [0, t_c] \times \mathcal{P} \longrightarrow \mathbb{R}^{n-m} \tag{6.4}$$

The trajectory compatibility property is therefore expressed as follows:

$$\forall p_1, p_2 \in \mathcal{P}, \begin{cases} T(0, p_1) = T(t_c, p_1) = T(0, p_2) = T(t_c, p_2) \\ \frac{d}{dt}T(0, p_1) = \frac{d}{dt}T(t_c, p_1) = \frac{d}{dt}T(0, p_2) = \frac{d}{dt}T(t_c, p_2) \end{cases} \tag{6.5}$$

The parameterization of the set of reference trajectories must be such that, for a given trajectory, we can evaluate the stability of the internal dynamics during the tracking of the references trajectories on the linearized part. Fig. 6.2 illustrates the set of parameterized trajectories.

### 6.2.3 Proposed control law

Under the condition of strong inertial coupling between actuated and non-actuated coordinates defined above, it is possible to linearize the non-actuated part of the system. Eq. (6.2) is solved for $\ddot{q}_a$ based on the pseudo-inverse of the matrix $M_{12}$:

$$\ddot{q}_a = -M_{12}^{\dagger}\left(M_{22}\ddot{q}_{na} + h_2 + g_2\right) \tag{6.6}$$

with $M_{12}^{\dagger}$, being the pseudo-inverse of the matrix $M_{12}$ given by:

$$M_{12}^{\dagger} = \left(M_{12}^{T}M_{12}\right)^{-1} M_{12}^{T} \tag{6.7}$$

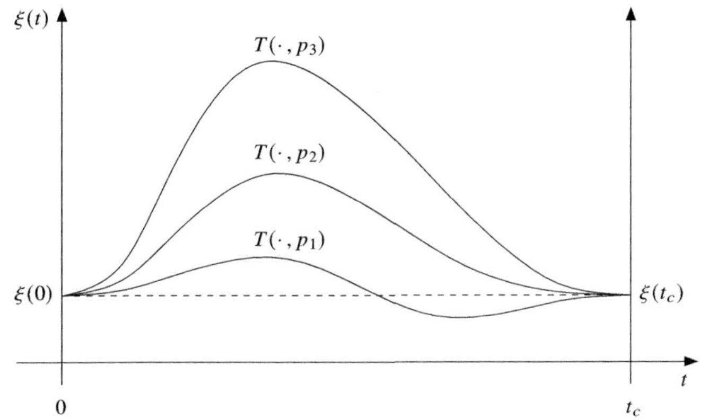


**FIGURE 6.2** A set of parametric reference trajectories to be tracked by linearized coordinates over a period of time $t_c$. The term $\xi = \left[q_{na}^T \; \dot{q}_{na}^T\right]^T$ is the state of the linearized part.

After replacing $\ddot{q}_a$ with (6.6), Eq. (6.1) becomes:

$$\left(M_{12} - M_{11}M_{12}^{\dagger}M_{22}\right)\ddot{q}_{na} + h_1 - M_{11}M_{12}^{\dagger}h_2 + g_1 - M_{11}M_{12}^{\dagger}g_2 = u \quad (6.8)$$

Let us now consider the following state feedback:

$$u = \left(M_{12} - M_{11}M_{12}^{\dagger}M_{22}\right)v + h_1 - M_{11}M_{12}^{\dagger}h_2 + g_1 - M_{11}M_{12}^{\dagger}g_2 \quad (6.9)$$

with $v$ an auxiliary control input.

The closed-loop dynamics of the system (6.2) and (6.1) is then equivalent to the following system:

$$\ddot{q}_{na} = v \quad (6.10)$$

$$\ddot{q}_a = -M_{12}^{\dagger}\left(M_{22}v + h_2 + g_2\right) \quad (6.11)$$

The auxiliary control input $v$ is used to track reference trajectories $T(t, p)$. The reference trajectories and their derivatives are denoted:

$$q_{na}^d(t) = T(t, p) \quad (6.12)$$

$$\dot{q}_{na}^d(t) = \frac{d}{dt}T(t, p) \quad (6.13)$$

$$\ddot{q}_{na}^d(t) = \frac{d^2}{dt^2}T(t, p) \quad (6.14)$$

The auxiliary command $v$ is then defined as follows (omitting time dependency $t$):

$$v = \ddot{q}_{na}^d + K_d\left(\dot{q}_{na}^d - \dot{q}_{na}\right) + K_p\left(q_{na}^d - q_{na}\right) \quad (6.15)$$

with $K_p$ and $K_d$ the positive-definite feedback gain matrices. The linearized sub-system (6.10) becomes, after injection of this additional command (6.15), equivalent to the following linear system:

$$\ddot{e} + K_d\dot{e} + K_p e = 0 \tag{6.16}$$

with the change of variables $e = q_{na}^d - q_{na}$, $\dot{e} = \dot{q}_{na}^d - \dot{q}_{na}$ and $\ddot{e} = \ddot{q}_{na}^d - \ddot{q}_{na}$.

The resulting system is an asymptotically stable linear system with positive-definite gain matrices $K_p$ and $K_v$ [20].

Let $z = \begin{bmatrix} e \\ \dot{e} \end{bmatrix}$ be the state vector. Eq. (6.16) can be written as an equation of state as follows:

$$\dot{z} = Az \tag{6.17}$$

with

$$A = \begin{bmatrix} 0 & I \\ -K_p & -K_d \end{bmatrix} \tag{6.18}$$

If $K_p$ and $K_d$ are positive, then $A$ is hurwitz (i.e., the real parts of $A$'s eigenvalues are strictly negative) and so this system is asymptotically stable. The system therefore tends asymptotically towards its equilibrium point at the origin, that is:

$$\lim_{t\to\infty} e(t) = \lim_{t\to\infty} \dot{e}(t) = 0 \tag{6.19}$$

As a result, the position and velocity of the linearized passive coordinates tend asymptotically towards the values of the reference trajectories, that is:

$$\lim_{t\to\infty} q_{na}(t) = q_{na}^d(t)$$
$$\lim_{t\to\infty} \dot{q}_{na}(t) = \dot{q}_{na}^d(t)$$

If these reference trajectories are periodic, the linearized part of the system joins a stable limit cycle [21–26].

Let us consider the internal dynamics of the sub-system expressed by Eq. (6.11) into which we inject the additional control law (6.15):

$$\ddot{q}_a = -M_{12}^{\dagger}\left(M_{22}\left[\ddot{q}_{na}^d + K_d\left(\dot{q}_{na}^d - \dot{q}_{na}\right) + K_p\left(q_{na}^d - q_{na}\right)\right] + h_2 + g_2\right) \tag{6.20}$$

This dynamics is stable if the system under consideration is minimum phase. In the more interesting case of non-minimum-phase systems [12–15], the internal dynamics is unstable and should be stabilized to ensure the stability of the system as a whole.

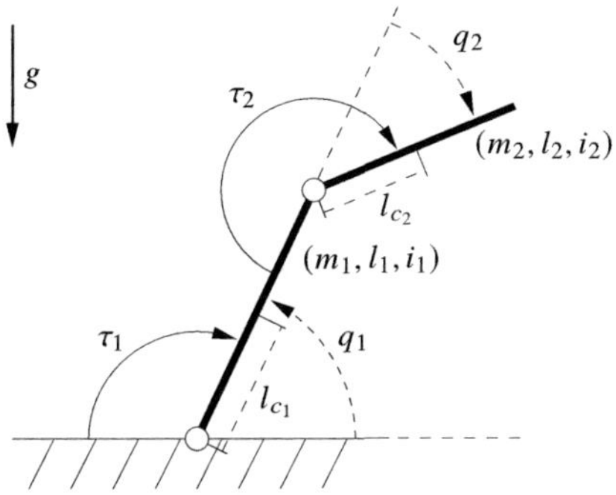


**FIGURE 6.3** Diagram of the double inverted pendulum. In the literature, this system is referred to as a *pendubot* if the actuator is located at $q_1$ or an *acrobot* if it is located at $q_2$.

To illustrate the principle of partial feedback linearization [9–11,18] and show the internal dynamics of an underactuated mechanical system, the proposed control approach will be applied in the following to an example of the double inverted pendulum.

### 6.2.4 An illustrative example

The principle of partial linearization by state feedback [9–11,18] is applied, by way of illustration, to a particular underactuated mechanical system: the inverted double pendulum.

Consider a kinematic chain made up of two solid bodies connected by a pivot link. The first solid body is connected to a fixed support by a pivot link running in the same plane as the first. This mechanical system is underactuated if it has only one actuator. We call $q = [q_1\ q_2]^T$ the generalized coordinate vector. In the literature, this underactuated mechanical system is called a *pendubot* if the actuator is located at the fixed support, or an *acrobot* if it is located at the link between the two solids. Fig. 6.3 shows the inverted double pendulum.

#### 6.2.4.1 Modeling

We recall the equations of the dynamics of this underactuated system obtained by the Euler–Lagrange formulation ([18]):

$$m_{11}\ddot{q}_1 + m_{12}\ddot{q}_2 + h_1 + g_1 = u_1 \tag{6.21}$$

$$m_{21}\ddot{q}_1 + m_{22}\ddot{q}_2 + h_2 + g_2 = u_2 \tag{6.22}$$

with

$$m_{11} = m_1 l_{c_1}^2 + m_2\left(l_1^2 + l_{c_2}^2 + 2 l_1 l_{c_2} \cos q_2\right) + i_1 + i_2$$

$$m_{22} = m_2 l_{c_2}^2 + i_2$$

$$
\begin{aligned}
m_{12} &= m_{21} = m_2\left(l_{c_2}^2 + l_1 l_{c_2}\cos q_2\right) + i_2 \\
h_1 &= -m_2 l_1 l_{c_2}\sin q_2\left(\dot{q}_2^2 + 2\dot{q}_1\dot{q}_2\right) \\
h_2 &= m_2 l_1 l_{c_2}\sin q_2 \dot{q}_1^2 \\
g_1 &= \left(m_1 l_{c_1} + m_2 l_1\right) g\cos q_1 + m_2 l_{c_2} g\cos\left(q_1+q_2\right) \\
g_2 &= m_2 l_{c_2} g\cos\left(q_1+q_2\right)
\end{aligned}
\tag{6.23}
$$

where $l_1$ and $l_2$ are the respective lengths of the two solids, $l_{c_1}$ and $l_{c_2}$ are the distances between the attachment to the preceding solid and the center of gravity, $m_1$ and $m_2$ are their respective masses, $i_1$ and $i_2$ their respective moments of inertia, and $\tau_1$ and $\tau_2$ the generalized external forces. In the case of the *acrobot* $u_1 = 0$ and in the case of the *pendubot* $u_2 = 0$.

#### 6.2.4.2 Partial feedback linearization

The principle of partial feedback linearization [9–11,18] is applied to both variants of the double inverted pendulum. In both cases, the non-actuated coordinate is linearized. Consider the following two state feedbacks:

$$
u_1 = \left(m_{12} - \frac{m_{11}m_{22}}{m_{12}}\right)v_1 + h_1 - \frac{m_{11}}{m_{12}}h_2 + g_1 - \frac{m_{11}}{m_{12}}g_2 \tag{6.24}
$$

$$
u_2 = \left(m_{12} - \frac{m_{22}m_{11}}{m_{12}}\right)v_2 + h_2 - \frac{m_{22}}{m_{12}}h_1 + g_2 - \frac{m_{22}}{m_{12}}g_1 \tag{6.25}
$$

with $v_1$ and $v_2$ as auxiliary control inputs. These control laws are defined if $m_{12} \neq 0, \forall q_1, q_2$, that is:

$$
\forall q_2\,,\quad m_2\left(l_{c_2}^2 + l_1 l_{c_2}\cos q_2\right) + i_2 \neq 0 \Longleftrightarrow \forall q_2\,,\quad \cos q_2 \neq -\frac{i_2 + m_2 l_{c_2}^2}{m_2 l_1 l_{c_2}} \tag{6.26}
$$

which is verified if

$$
\frac{i_2 + m_2 l_{c_2}^2}{m_2 l_1 l_{c_2}} > 1 \tag{6.27}
$$

that is $i_2 > m_2 l_{c_2}(l_1 - l_{c_2})$. The closed-loop system is then equivalent to the following system for the acrobot:

$$
\ddot{q}_1 = v_1 \tag{6.28}
$$

$$
\ddot{q}_2 = -\frac{1}{m_{12}}(m_{11} + h_1 + g_1)v_1 \tag{6.29}
$$

and the following system for the pendubot:

$$
\ddot{q}_1 = -\frac{1}{m_{12}}(m_{22} + h_2 + g_2)v_2 \tag{6.30}
$$

$$\ddot{q}_2 = v_2 \tag{6.31}$$

#### 6.2.4.3 Control law

Given the reference trajectories $q_1^d(t)$ for the acrobot and $q_2^d(t)$ for the pendubot, we choose the auxiliary control input:

$$v_1 = \ddot{q}_1^d + k_{1d}\left(\dot{q}_1^d - \dot{q}_1\right) + k_{1p}\left(q_1^d - q_1\right) \tag{6.32}$$

$$v_2 = \ddot{q}_2^d + k_{2d}\left(\dot{q}_2^d - \dot{q}_2\right) + k_{2p}\left(q_2^d - q_2\right) \tag{6.33}$$

with $k_{1d}, k_{2d}, k_{1p}, k_{2p}$ positive feedback gains.

#### 6.2.4.4 Linearized dynamics

The dynamics of the linearized closed-loop part is then expressed after a change of variables as: $e_1 = \left(q_1^d - q_1\right)$, $\dot{e}_1 = \left(\dot{q}_1^d - \dot{q}_1\right)$, and $\ddot{e}_1 = \left(\ddot{q}_1^d - \ddot{q}_1\right)$ for the acrobot, and $e_2 = \left(q_2^d - q_2\right)$, $\dot{e}_2 = \left(\dot{q}_2^d - \dot{q}_2\right)$, and $\ddot{e}_2 = \left(\ddot{q}_2^d - \ddot{q}_2\right)$ for the pendubot:

$$\ddot{e}_1 + k_{1d}\dot{e}_1 + k_{1p}e_1 = 0 \tag{6.34}$$

$$\ddot{e}_2 + k_{2d}\dot{e}_2 + k_{2p}e_2 = 0 \tag{6.35}$$

The systems (6.34) and (6.35) are asymptotically stable if the feedback gains are positive. The non-actuated coordinate follows the reference trajectory.

#### 6.2.4.5 Internal dynamics

Let us now analyze the dynamics of the nonlinearized closed-loop part. The internal dynamics are obtained by replacing the auxiliary control laws (6.32) and (6.33) in the equations of the dynamics of the actuated part in a closed loop, respectively, (6.29) and (6.30):

$$\ddot{q}_2 = -\frac{h_1 + g_1}{m_{12}} - \frac{m_{11}}{m_{12}}\left[\ddot{q}_1^d + k_{1d}\left(\dot{q}_1^d - \dot{q}_1\right) + k_{1p}\left(q_1^d - q_1\right)\right] \tag{6.36}$$

$$\ddot{q}_1 = -\frac{h_2 + g_2}{m_{12}} - \frac{m_{22}}{m_{12}}\left[\ddot{q}_2^d + k_{2d}\left(\dot{q}_2^d - \dot{q}_2\right) + k_{2p}\left(q_2^d - q_2\right)\right] \tag{6.37}$$

Consider the special case of stabilization around the unstable equilibrium point $(q_1, q_2) = \left(\frac{\pi}{2}, 0\right)$. We wish to control the pendulum to stabilize the high unstable equilibrium position $(q_1, q_2) = \left(\frac{\pi}{2}, 0\right)$. We then assume that $q_1^d = \frac{\pi}{2}$, $\dot{q}_1^d = \ddot{q}_1^d = 0$ for the acrobot, and $q_2^d = \dot{q}_2^d = \ddot{q}_2^d = 0$ for the pendubot. Perfect tracking of the reference trajectory is also assumed. By making these substitutions in the equations of the internal dynamics, we obtain the following equations describing the dynamics of the zeros of the acrobot and pendubot, respectively:

$$\left(l_{c_2} + l_1 \cos q_2 + \frac{i_2}{m_2 l_{c_2}}\right)\ddot{q}_2 = \left(l_1 \dot{q}_2^2 - g\right)\sin q_2 \tag{6.38}$$

**TABLE 6.1** Summary of the dynamic parameters of the double inverted pendulum used in simulations.

| Parameter | Value | Unit |
|---|---|---|
| $m_1$ | 1 | kg |
| $m_2$ | 1 | kg |
| $l_1$ | 1 | m |
| $l_2$ | 1 | m |
| $l_{c1}$ | 0.5 | m |
| $l_{c2}$ | 0.5 | m |
| $i_1$ | 0.2 | kg m$^2$ |
| $i_2$ | 1 | kg m$^2$ |
| $g$ | 9.8 | m s$^{-2}$ |

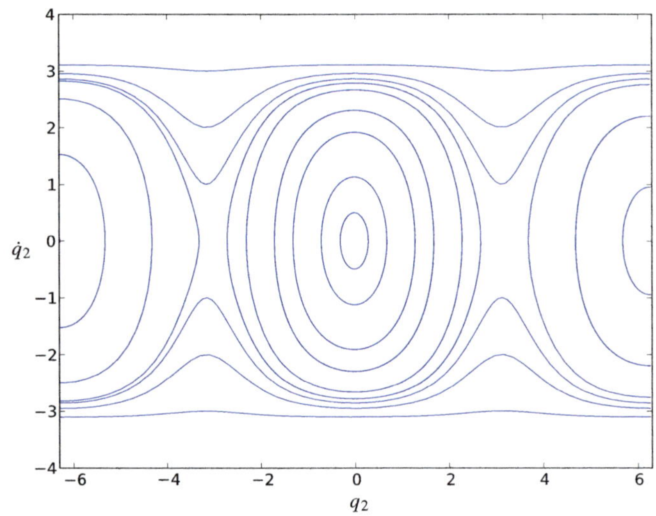


**FIGURE 6.4** Phase portrait of the zero dynamics of the acrobot.

$$\left(l_{c_2} + l_1 + \frac{i_2}{m_2 l_{c_2}}\right)\ddot{q}_1 = -g\cos q_1 \tag{6.39}$$

Fig. 6.4 represents the phase portrait of the acrobot's zero dynamics described by Eq. (6.38). Fig. 6.5 represents the pendubot dynamics described by Eq. (6.39). These figures are obtained with the system dynamic parameters described in Table 6.1. The system described by Eq. (6.38) whose phase portrait is represented by Fig. 6.4 has two equilibrium points: $(q_2, \dot{q}_2) = (0, 0)$, unstable (saddle point) and $(q_2, \dot{q}_2) = (\pi, 0)$, stable (*center point*). The system described by Eq. (6.39)

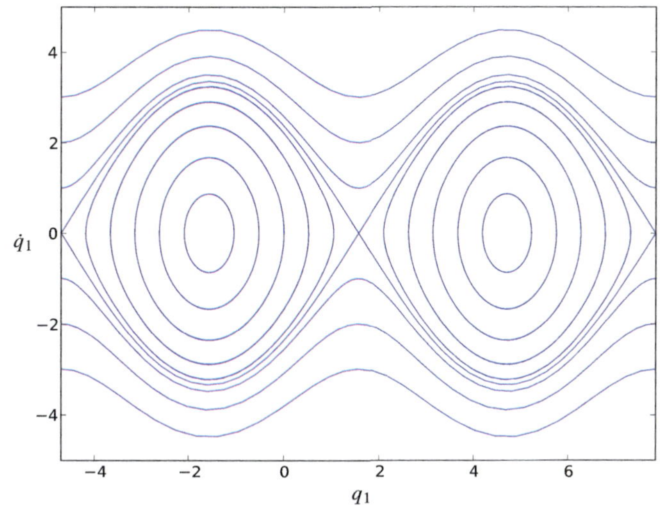


**FIGURE 6.5** Phase portrait of the zero dynamics of the pendubot.

whose phase plane is represented by Fig. 6.5 also has two equilibrium points: $(q_1, \dot{q}_1) = \left(\frac{\pi}{2}, 0\right)$, unstable and $(q_1, \dot{q}_1) = \left(-\frac{\pi}{2}, 0\right)$, stable.

## 6.3 Stabilization of the internal dynamics

The control law is completely defined given the reference trajectories $q_i^d(t)$, $\dot{q}_i^d(t)$, $\ddot{q}_i^d(t)$. The control approach described in this chapter consists of optimizing the reference trajectories in order to stabilize the system's internal dynamics.

### 6.3.1 Optimization of reference trajectories

The aim of optimizing reference trajectories is to stabilize the system's internal dynamics, generating a stable limit cycle [21–26] on the nonlinear part of the system (internal dynamics), while maintaining the stability of the linearized part around a stable limit cycle (obtained through the periodic trajectories to be followed). Reference trajectories are parameterized in the trajectory set according to their influence on an internal dynamics stability criterion. The parameter of the reference trajectories is used as an auxiliary control input to stabilize the internal dynamics. At the beginning of each cycle, given the system configuration, the optimal trajectory in terms of this stability criterion is calculated and tracked for the duration of the whole cycle. The parameter is chosen to minimize the divergence of the nonlinearized coordinate trajectories. This operation is repeated at the beginning of each cycle. Thanks to this optimization, the internal dynamics of the system, made up of the actuated coordinates, is stabilized. As a result of the periodicity of the reference trajectories, the trajectories of the actuated coordinates are maintained around a limit cycle. As a result, all coordinates (actuated and unactuated) converge towards stable limit cycles [21–26].

The cost function $\mathcal{F}$ is defined as a positive scalar function of the system state at the beginning of the $k$th cycle and of the reference trajectory optimization parameter:

$$\begin{array}{lrcl} \mathcal{F}: & \mathbb{R}^n \times \mathbb{R}^n \times \mathcal{P} & \longrightarrow & \mathbb{R} \\ & (q((k-1)t_c), \dot{q}((k-1)t_c), p) & \longmapsto & \mathcal{F}\left(q((k-1)t_c), \dot{q}((k-1)t_c), p\right) \end{array} \tag{6.40}$$

The value of the cost function $\mathcal{F}$ is directly related to the state of the system at the end of the cycle $(q\,(kt_c)\,,\dot{q}\,(kt_c))$ so that if $\mathcal{F}\,(q((k-1)t_c), \dot{q}((k-1)t_c), p)$ $= 0$ then the system joins a stable limit cycle.

At the beginning of cycle $k$, given the system state $(q(kt_c), \dot{q}(kt_c))$, the optimal parameter $p^{opt}$ is calculated by solving the following optimization problem:

$$p^{opt} = \arg\min_{p \in \mathcal{P}} \mathcal{F}\left(q((k-1)t_c), \dot{q}((k-1)t_c), p\right) \tag{6.41}$$

The $p^{opt}$ parameter calculated in this way will be used to completely determine the reference trajectories used to control the system for the entire $k$ cycle:

$$q_{na}^{d}(t) = T(t, p^{opt}) \tag{6.42}$$

$$\dot{q}_{na}^{d}(t) = \frac{d}{dt} T(t, p^{opt}) \tag{6.43}$$

$$\ddot{q}_{na}^{d}(t) = \frac{d^2}{dt^2} T(t, p^{opt}) \tag{6.44}$$

### 6.3.2 Estimation and external disturbance rejection

The optimization of reference trajectories makes it possible to reject punctual disturbances. This category of disturbance refers to those whose intensity changes over time. Random phenomena, for example, or the application of an external force to the system not foreseen in the model, are point disturbances. No model can predict the full range of events independent of the system that may influence its dynamics.

The $\mathcal{F}$ function that translates the cost function uses a nominal model of the system, taking no external disturbances into account. One-off disturbances are identified at the end of a cycle and rejected if their effect on system dynamics does not persist into the next cycle.

Certain types of disturbance can have permanent effects on the system. They are referred to as permanent or persistent disturbances. Persistent disturbances have the particularity of influencing system dynamics continuously and with the same intensity, even though they are not foreseen in the model. Mechanical faults, for example, can induce persistent disturbances. Errors in estimating the physical parameters of a model can also induce persistent disturbances. For example, an error in estimating the mass of a mechanical part can lead to uncertainty, resulting in a persistent disturbance.

This type of disturbance needs to be treated in a specific way. The principle implemented in our case is to highlight the effects of these disturbances on the state of the system at the end of a reference trajectory continuation cycle. Indeed, it is possible to take advantage of the periodicity of reference trajectories to construct a nonlinear estimator of persistent disturbances.

The effect of the disturbance is modeled as a constant shift in the position measurement of the system's coordinates:

$$q^{\star} = q + \epsilon \tag{6.45}$$

where $q^{\star}$ is the incorrectly measured value and $\epsilon$ is the measurement error. Thus for velocities, we have:

$$\dot{q}^{\star} = \dot{q} + \frac{d}{dt}\epsilon = \dot{q} \tag{6.46}$$

since $\epsilon$ is assumed constant. The method then consists in writing the state of the system $(q, \dot{q})$ at the end of the cycle, assuming that the tracking during the last cycle was biased by the wrong position measurement. We will then quantify the difference between the state of the system and an undisturbed nominal system as a function of the $\epsilon$ offset. The control law can then be modified to take account of this estimated offset, thus rejecting the persistent disturbance.

The linearizing control law $\tau\,(q, \dot{q})$ is examined, taking into account the fact that the calculations use the measured value $q^{\star}$. In order to write the distorted control law $\tau\,(q^{\star}, \dot{q})$ in terms of the control law that, virtually, would have access to the real value $q$, we linearize around $(q^{\star}, \dot{q})$:

$$\tau\left(q^{\star}, \dot{q}\right) = \tau\,(q + \epsilon, \dot{q}) \approx \tau\,(q, \dot{q}) + \epsilon\frac{\partial}{\partial q}\tau\,(q, \dot{q}) \tag{6.47}$$

We now examine the closed-loop system using this control law. We take up the equations of the dynamics of the partially linearized system (6.10) and (6.11), taking into account Eq. (6.47):

$$\ddot{q}_{na} = v + \epsilon\frac{\partial}{\partial q}\tau\,(q, \dot{q}) \tag{6.48}$$

$$\ddot{q}_a = \ddot{q}_a^{nom} - M_{12}^{\dagger}M_{22}\epsilon\frac{\partial}{\partial q}\tau\,(q, \dot{q}) \tag{6.49}$$

where $\ddot{q}_a^{nom}$ denotes the acceleration of the $q_a$ coordinate in the case of the nominal, undisturbed system.

In order to observe the state of the system at the end of the tracking cycle, assuming that the tracking has been carried out according to the control law (6.47), we integrate the dynamics over a complete period (one cycle), which gives:

$$\int_0^{t_c} \ddot{q}_a(t)dt = \int_0^{t_c} \ddot{q}_a^{nom}(t)dt - \int_0^{t_c} M_{12}^{\dagger} M_{22} \epsilon \frac{\partial}{\partial q} \tau(q, \dot{q})\, dt$$

$$\dot{q}_a(t_c) - \dot{q}_a(0) = \dot{q}_a^{nom}(t_c) - \dot{q}_a^{nom}(0) - \int_0^{t_c} M_{12}^{\dagger} M_{22} \epsilon \frac{\partial}{\partial q} \tau(q, \dot{q})\, dt \quad (6.50)$$

The control law is corrected iteratively after each period by integrating the offset (due to the disturbance) estimated in the previous period. Under these conditions, we can assume that the system at the beginning of the cycle is identical to the nominal system, and therefore that $\dot{q}_a(0) = \dot{q}_a^{nom}(0)$. We then have:

$$\dot{q}_a(t_c) - \dot{q}_a^{nom}(t_c) = -\int_0^{t_c} M_{12}^{\dagger} M_{22} \epsilon \frac{\partial}{\partial q} \tau(q, \dot{q})\, dt \quad (6.51)$$

We have quantified the error on the velocity of the actuated coordinates (zero dynamics) caused by a control law biased by a $\epsilon$ measurement error. If the integral of Eq. (6.51) cannot be calculated exactly, it can be calculated numerically over a cycle during trajectory tracking.

Note that calculating the error on the state of the system at the end of the cycle requires exact knowledge of the state $(q, \dot{q})$, but the control law only has access to the biased state $(q^\star, \dot{q})$. Considering Eq. (6.45), we linearize the term dependent on $(q, \dot{q})$ around $q^\star$:

$$\begin{aligned} \frac{\partial}{\partial q} \tau(q, \dot{q}) &= \frac{\partial}{\partial q} \tau\left(q^\star - \epsilon, \dot{q}\right) \\ &= \frac{\partial}{\partial q} \tau\left(q^\star, \dot{q}\right) + \epsilon \frac{\partial^2}{\partial q^2} \tau\left(q^\star, \dot{q}\right) \end{aligned} \quad (6.52)$$

Eq. (6.51) then becomes:

$$\dot{q}_a(t_c) - \dot{q}_a^{nom}(t_c) = -\int_0^{t_c} M_{12}^{\dagger} M_{22} \epsilon \left[\frac{\partial}{\partial q} \tau\left(q^\star, \dot{q}\right) + \epsilon \frac{\partial^2}{\partial q^2} \tau\left(q^\star, \dot{q}\right)\right] \quad (6.53)$$

which can be seen as a polynomial of degree two:

$$Q(\epsilon) = a\left(q^\star, \dot{q}\right)\epsilon^2 + b\left(q^\star, \dot{q}\right)\epsilon + c\left(q^\star, \dot{q}\right) = 0 \quad (6.54)$$

where $a$, $b$, and $c$ are coefficients dependent on $(q^\star, \dot{q})$ and $\dot{q}_a(t_c) - \dot{q}_a^{nom}(t_c)$.

The roots of this polynomial can be calculated online, at the end of each cycle. If the polynomial has no roots, we choose the value of $\epsilon$ that minimizes $\|Q(\epsilon)\|$. The disturbance estimation is updated at the end of each period (cycle) as follows:

$$\begin{cases} \epsilon_{k+1} = \epsilon_k + \Delta\epsilon_{k+1} \\ \epsilon_0 = 0 \end{cases} \quad (6.55)$$

The value $\Delta\epsilon_{k+1}$ is the root calculated at the end of period $k$, given the values observed at the end of period $k$, assuming that the control law calculation during period $k$ uses the adjusted value $q^\star - \epsilon_k$.

## 6.4 Application: inertia wheel inverted pendulum

In this section, the proposed control scheme detailed in previous sections will be applied to the case of an inertia wheel inverted pendulum.

### 6.4.1 Reference trajectories generation

The reference trajectories are designed to satisfy the criteria imposed on the desired dynamics of the inverted pendulum. Their shape must satisfy various constraints. These are consequences of the laws of physics, as well as those imposed by the control designer. Several considerations must therefore be taken into account.

The reference trajectory $\theta_1^{ref}(t)$ must be of class $C^2$, i.e., twice continuously differentiable. This ensures that the trajectory has no discontinuities in position, velocity, or acceleration. The need for continuity in position is obvious: a discontinuity in position translates into a jump from one position to another in an infinitely short time. Continuity in velocity translates into the absence of shocks that could damage the equipment. Finally, continuity in acceleration is imposed by the motor's internal mechanisms that do not allow an instantaneous change in the value of the torque it generates.

The desired behavior of the pendulum is a periodic motion symmetrical with respect to the point of unstable equilibrium. The reference trajectory for the non-actuated coordinate $\theta_1$ must therefore be a symmetrical periodic trajectory.

In the real case of a robot subject to disturbances, both from external sources and inherent in the model, which can only be approximated to a certain degree, the tracking of a symmetrical reference trajectory on the $\theta_1$ coordinate of the pendulum leads to an aperiodic motion of the flywheel, despite the structural symmetry of the system's dynamics. The result is a non-zero flywheel velocity at the end of each cycle. For this reason, the originally symmetrical reference trajectories to be followed by the pendulum must be able to be deformed to force the pendulum to spend more time on one side of the vertical axis than on the other one. In this way, it is possible to correct the consequences of disturbances on the flywheel's speed by choosing a suitable deformation, and thus achieve zero speed at the end of the cycle.

Reference trajectories satisfying the above constraints are designed as follows. A complete cycle is divided into two half-periods. Each part takes the form of a parameterized polynomial function. The reference trajectory over a complete cycle is then expressed as follows:

$$\theta_1^{ref}(t,p) = \begin{cases} AP(\frac{2t}{t_c}, p) & \text{si } t \in [0, \frac{t_c}{2}[ \\ -AP(\frac{2t}{t_c} - 1, 1-p) & \text{si } t \in [\frac{t_c}{2}, t_c[ \end{cases} \tag{6.56}$$

with $A$ and $t_c$, respectively, the amplitude and period of the oscillations and $p$ the optimization parameter that will deform the trajectory.

$P(t, p)$ is the polynomial function in $t$ that varies from 1 at time $t = 0$ to $-1$ at time $t = 1$, for all $p$ in an interval $[p_{min}, p_{max}]$. To obtain smooth junctions (continuity constraints) between successive parts of the trajectory, the first and second derivatives of $P(t, p)$ with respect to time are constrained to the value 0 at the extremities (i.e., at instants $t = 0$ and $t = 1$). The parameter $p$ defines the normalized time at which the pendulum reaches the vertical position, i.e., $P(t = p, p) = 0$. The range of admissible values of the optimization parameter $[p_{min}, p_{max}]$ is a sub-interval of $]0, 1[$. The trajectory is therefore symmetrical with respect to the vertical when $p = \frac{1}{2}$.

These considerations lead us to formally establish this set of constraints:

$$\begin{cases} P(0,p) = 1 \\ P(1,p) = -1 \\ \frac{\partial P}{\partial t}(0,p) = 0 \\ \frac{\partial P}{\partial t}(1,p) = 0 \\ \frac{\partial^2 P}{\partial t^2}(0,p) = 0 \\ \frac{\partial^2 P}{\partial t^2}(1,p) = 0 \\ P(t=p,p) = 0 \end{cases} \tag{6.57}$$

These seven constraints can be satisfied if the parameterized polynomial $P(t, p)$ has a degree of at least six. This polynomial is then defined as follows:

$$P(t,p) = \sum_{i=0}^{6} \alpha_i(p) t^i \tag{6.58}$$

The coefficients of this polynomial are the solutions to the following problem, written in matrix form:

$$\begin{bmatrix} 0 & 0 & 0 & 0 & 0 & 0 & 1 \\ 1 & 1 & 1 & 1 & 1 & 1 & 1 \\ 0 & 0 & 0 & 0 & 0 & 1 & 0 \\ 6 & 5 & 4 & 3 & 2 & 1 & 0 \\ 0 & 0 & 0 & 0 & 1 & 0 & 0 \\ 30 & 20 & 12 & 6 & 2 & 0 & 0 \\ p^6 & p^5 & p^4 & p^3 & p^2 & p & 1 \end{bmatrix} \begin{bmatrix} \alpha_6 \\ \alpha_5 \\ \alpha_4 \\ \alpha_3 \\ \alpha_2 \\ \alpha_1 \\ \alpha_0 \end{bmatrix} = \begin{bmatrix} 1 \\ -1 \\ 0 \\ 0 \\ 0 \\ 0 \\ 0 \end{bmatrix} \tag{6.59}$$

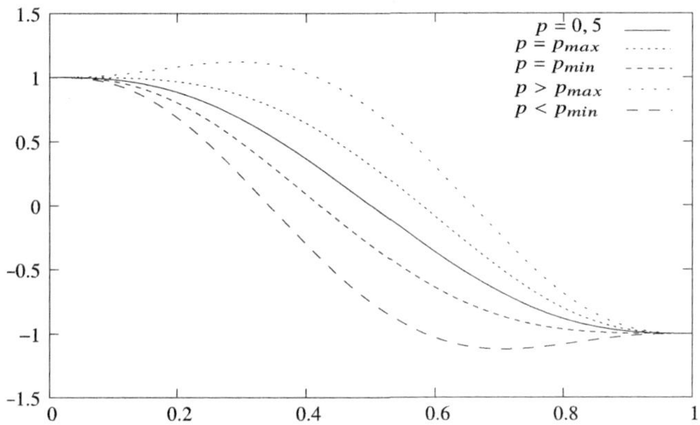


**FIGURE 6.6** Evolution of the normalized reference trajectory of $\theta_1$ over a half- period for different values of $p$.

Solving the matrix Eq. (6.59) yields the following $\alpha_i(p)$ coefficients:

$$\begin{aligned}
\alpha_6(p) &= \frac{(2p-1)(6p^4-12p^3+4p^2+2p+1)}{p^3(p-1)^3} \\
\alpha_5(p) &= \frac{-3(4p^6-18p^4+16p^3-1)}{p^3(p-1)^3} \\
\alpha_4(p) &= \frac{3(10p^6-18p^5+10p^3-1)}{p^3(p-1)^3} \\
\alpha_3(p) &= \frac{-(20p^6-48p^5+30p^4-1)}{p^3(p-1)^3} \\
\alpha_2(p) &= 0 \\
\alpha_1(p) &= 0 \\
\alpha_0(p) &= 1
\end{aligned} \tag{6.60}$$

The parameter $p$ is by definition within a sub-interval of $]0, 1[$, so the coefficients are always well defined. To illustrate the influence of parameter $p$ on the shape of the resulting trajectory, Fig. 6.6 shows the evolution of this trajectory for different values of $p$. The pendulum spends as much time on either side of the unstable equilibrium point (i.e., the vertical) for $p = \frac{1}{2}$.

To prevent the pendulum body from touching the mechanical stops and to respect an oscillatory motion, the pendulum angle is constrained to remain within the interval $[-A, A]$. This is why the range of permissible values for parameter $p$ is restricted to a sub-interval $[p_{min}, p_{max}]$ of $]0, 1[$. The formalization of these constraints can be expressed as follows, for normalized trajectories:

$$\forall p \in [p_{min}, p_{max}], \forall t \in [0, 1], |P(t, p)| \leq 1 \tag{6.61}$$

where $p_{min}$ and $p_{max}$ are the extreme values, calculated as follows.

Suppose that for a certain $p$, we have:

$$\exists t_1 \in [0, 1] \quad \text{such as} \quad |P(t_1, p)| > 1 \tag{6.62}$$

Consider the three propositions $P(0, p) = 1$, $P(1, p) = -1$, and (6.62), $P$, being continuous and continuously differentiable, it must admit at least one point where its tangent is zero. In other words, at least one root of the polynomial $\frac{\partial P(t,p)}{\partial t}$ lies in the interval [0, 1]. Calculating the real roots of $\frac{\partial P(t,p)}{\partial t}$ leads to the unique solution:

$$\left.\frac{\partial P(t, p)}{\partial t}\right|_{t=t_1} = 0 \quad \Rightarrow \quad t_1 = \frac{20\,p^6 - 48\,p^5 + 30\,p^4 - 1}{24\,p^5 - 60\,p^4 + 40\,p^3 - 2} \tag{6.63}$$

The extreme trajectories $P(t, p_{min})$ and $P(t, p_{max})$ correspond, respectively, to the 0 and 1 values of $t_1$. The extreme values of the parameter $p$ are therefore the real solutions in the interval [0, 1] of the following two equations:

$$\frac{20\,p^6 - 48\,p^5 + 30\,p^4 - 1}{24\,p^5 - 60\,p^4 + 40\,p^3 - 2} = 0 \tag{6.64}$$

$$\frac{20\,p^6 - 48\,p^5 + 30\,p^4 - 1}{24\,p^5 - 60\,p^4 + 40\,p^3 - 2} = 1 \tag{6.65}$$

Eq. (6.64) admits a unique real solution $p = p_{min}$, Eq. (6.65) has $p = p_{max}$ as its unique real solution. Both solutions lie in the interval [0, 1].

**Remark 2.** The numerical solutions of the two Eqs. (6.64) and (6.65) are, respectively, $p_{min} \approx 0.42$ and $p_{max} \approx 0.58$. Note that the complementarity between $p_{min}$ and $p_{max}$ results in $p_{min} = 1 - p_{max}$.

### 6.4.2 Control law

The dynamic model of the inertia wheel inverted pendulum can be rewritten in this form, isolating the acceleration vector $[\ddot{\theta}_1\ \ddot{\theta}_2]^T$:

$$\begin{bmatrix} \ddot{\theta}_1 \\ \ddot{\theta}_2 \end{bmatrix} = \begin{bmatrix} \frac{1}{I} & -\frac{1}{I} \\ -\frac{1}{I} & \frac{I+i_2}{Ii_2} \end{bmatrix} \left( \begin{bmatrix} C_1 \\ C_2 \end{bmatrix} - \begin{bmatrix} -\overline{ml}g\sin\theta_1 \\ 0 \end{bmatrix} \right) \tag{6.66}$$

The matrix Eq. (6.66) is equivalent to the following system of equations:

$$\ddot{\theta}_1 = \frac{1}{I}\left(C_1 - C_2 + \overline{ml}g\sin\theta_1\right) \tag{6.67a}$$

$$\ddot{\theta}_2 = -\frac{1}{I}C_1 + \frac{I+i_2}{Ii_2}C_2 - \frac{\overline{ml}g}{I}\sin\theta_1 \tag{6.67b}$$

These equations describe the evolution of the input–output dynamics of the open-loop system. According to the state feedback linearization method described above, we consider the following control law:

$$\tau = \overline{ml}g\sin\theta_1 - I\left[\ddot{\theta}_1^{ref} - k_p\left(\theta_1 - \theta_1^{ref}\right) - k_v\left(\dot{\theta}_1 - \dot{\theta}_1^{ref}\right)\right] \tag{6.68}$$

where $k_p$ and $k_v$ are positive constants corresponding, respectively, to the proportional and derivative feedback gains. These parameters are used to adjust the response of the closed-loop system.

In analyzing the dynamics of the closed-loop system, we assume that external disturbances (disturbance torques) are zero. The set of forces $[C_1\ C_2]^T$ is therefore reduced to the torque generated by the actuator motor alone. Therefore:

$$\begin{bmatrix} C_1 \\ C_2 \end{bmatrix} = \begin{bmatrix} 0 \\ \tau \end{bmatrix} \tag{6.69}$$

By substituting (6.68) and (6.69) into (6.67), we obtain the equations governing the dynamics of the closed-loop system:

$$\ddot{\theta}_1 = \ddot{\theta}_1^{ref} - k_p\left(\theta_1 - \theta_1^{ref}\right) - k_v\left(\dot{\theta}_1 - \dot{\theta}_1^{ref}\right) \tag{6.70a}$$

$$\ddot{\theta}_2 = \frac{\overline{ml}g}{i_2}\sin\theta_1 - \frac{I + i_2}{i_2}\left[\ddot{\theta}_1^{ref} - k_v\left(\dot{\theta}_1 - \dot{\theta}_1^{ref}\right) - k_p\left(\theta_1 - \theta_1^{ref}\right)\right] \tag{6.70b}$$

Consider the first sub-system corresponding to the linearized coordinate $\theta_1$, whose dynamics are described by Eq. (6.70a). By a change of variable, this subsystem is equivalent to the linear system described by the following equation:

$$\ddot{e} + k_v\dot{e} + k_p e = 0 \tag{6.71}$$

where $e = \theta_1 - \theta_1^{ref}$, $\dot{e} = \dot{\theta}_1 - \dot{\theta}_1^{ref}$, and $\ddot{e} = \ddot{\theta}_1 - \ddot{\theta}_1^{ref}$. The resulting system is an asymptotically stable linear system for any choice of positive gains $k_p$, $k_v$ [20]. The system therefore tends asymptotically towards its equilibrium point at the origin, i.e.:

$$\lim_{t\to\infty} e(t) = \lim_{t\to\infty} \dot{e}(t) = 0 \tag{6.72}$$

As a result, the position and velocity trajectories of the linearized $\theta_1$ coordinate converge to the reference trajectories:

$$\lim_{t\to\infty} \theta_1(t) = \theta_1^{ref}(t)$$

$$\lim_{t\to\infty} \dot{\theta}_1(t) = \dot{\theta}_1^{ref}(t)$$

In other words, the pendulum axis follows its reference trajectory. The dynamics described by Eq. (6.70b), on the other hand, are the internal dynamics of the system. These dynamics govern the evolution of the flywheel's motion over time, as the pendulum follows its reference trajectory. In order to stabilize these dynamics and obtain a global evolution of the inverted pendulum that satisfies the criteria we have set ourselves, the reference trajectories are optimized. In other words, the pendulum axis follows its reference trajectory.

The dynamics described by Eq. (6.70b), on the other hand, are the internal dynamics of the system. These dynamics govern the evolution of the flywheel's motion over time, as the pendulum follows its reference trajectory. In order to stabilize these dynamics and obtain a global evolution of the inverted pendulum that satisfies the criteria we have set ourselves, the reference trajectories are optimized.

### 6.4.3 Optimization of reference trajectories

The controller has a family of reference trajectories for the $\theta_1$ coordinate. As we saw in Section 6.4.2, this coordinate and its derivative asymptotically follow the reference trajectories $\theta_1^{ref}, \dot{\theta}_1^{ref}$. On the other hand, the reference trajectories satisfy the periodicity criterion, resulting in the generation of a stable limit cycle [21–26] on the non-actuated coordinate $\theta_1$. To stabilize the internal dynamics representing the nonlinearized active coordinate (the flywheel), the reference trajectory that minimizes the flywheel speed at the end of the cycle is chosen from an infinite number of trajectories obtained for different values of parameter $p$. More formally, given the flywheel speed at the start of cycle $\dot{\theta}_2(0)$, we need to find the parameter $p^{opt} \in [p_{min}, p_{max}]$ that minimizes the flywheel speed at the end of cycle $\dot{\theta}_2(t_c)$. The closed-loop dynamics of the flywheel are expressed by Eq. (6.70b). Let $g(p, \theta_1(t), \dot{\theta}_1(t), \dot{\theta}_2(t))$ be a quadratic function of the angular velocity $\dot{\theta}_2$ at the end of the cycle during trajectory tracking, given initial values $\theta_1(0)$, $\dot{\theta}_1(0)$, and $\dot{\theta}_2(0)$ and trajectory parameter $p$. The optimization problem then consists in finding the value $p^{opt}$ such that:

$$p^{opt} = \arg \min_{p \in [p_{min}, p_{max}]} g(p, \theta_1(t), \dot{\theta}_1(t), \dot{\theta}_2(t)) \tag{6.73}$$

The solution to this optimization problem is obtained by numerical integration of Eq. (6.70b). This optimization problem is solved at the beginning of each cycle. The optimal solution $p^{opt}$ allows us to completely define the reference trajectories to be followed throughout the cycle.

### 6.4.4 Estimation and rejection of persistent disturbances

The optimization of reference trajectories described in Section 6.4.3 naturally allows the rejection of point disturbances, and maintains the internal dynamics of the closed-loop system around a stable limit cycle [21–26]. However, this control method does not necessarily guarantee the rejection of persistent disturbances. Improving the ability to reject disturbances can be achieved by extending the control with a disturbance estimator that quantitatively assesses the consequence of the disturbance on the internal dynamics of the system. This estimator can be used to synthesize a control law that stabilizes the internal dynamics around a limit cycle despite the disturbance in question. The system is then better able to withstand the disturbance.

To enable the controller to reject these persistent disturbances without overloading it (to avoid actuator saturation), an iterative estimation of disturbances is performed at the end of each cycle. More precisely, the influence of these disturbances on the system's internal dynamics is estimated through observation of the flywheel speed, then translated into an additive term $\delta$ on the angular position $\theta_1$ of the pendulum. This term is then taken into account in the calculation of the control law expressed by Eq. (6.68). This results in a shift of the trajectory to one side of the unstable equilibrium position, which will compensate for the effect of the persistent disturbance.

To demonstrate the relationship between flywheel speed at the end of the cycle and the estimated $\delta$ shift, the effect of this disturbance is introduced into the model as follows:

$$\theta_1^* = \theta_1 + \delta \tag{6.74}$$

where $\theta_1^*$ is the actual disturbed position of the pendulum and $\theta_1$ is the position measured by the inclinometer. The disturbance parameter $\delta$ must be estimated.

The modified control law $\tau^*$ is obtained by taking into account the effects of the disturbance modeled by Eq. (6.74). The calculation of $\tau^*$ is the same as Eq. (6.68) but uses the estimated angular position of the pendulum $\theta_1^*$ instead of the measured value $\theta_1$:

$$\tau^* = \overline{ml}g \sin\theta_1^* - I\left[\ddot{\theta}_1^{ref} - k_p\left(\theta_1^* - \theta_1^{ref}\right) - k_v\left(\dot{\theta}_1^* - \dot{\theta}_1^{ref}\right)\right] \tag{6.75}$$

with $\dot{\theta}_1^* = \frac{d}{dt}\theta_1^* = \dot{\theta}_1 + \frac{d}{dt}\delta$. Since $\delta$ is assumed constant, we have $\frac{d}{dt}\delta = 0$ and therefore $\dot{\theta}_1^* = \dot{\theta}_1$. Substituting $\theta_1^*$ and $\dot{\theta}_1^*$ by their values in (6.75), we obtain:

$$\tau^* = \overline{ml}g \sin(\theta_1 + \delta) - I\left[\ddot{\theta}_1^{ref} - k_p\left(\theta_1 + \delta - \theta_1^{ref}\right) - k_v\left(\dot{\theta}_1 - \dot{\theta}_1^{ref}\right)\right] \tag{6.76}$$

We assume that $\delta$ is small enough to linearize the sinus in (6.76) without the approximation error affecting the efficiency of the approach:

$$\sin(\theta_1 + \delta) \approx \sin\theta_1 + \delta\cos\theta_1 \tag{6.77}$$

We can then express the modified control law $\tau^*$ in terms of the nominal one $\tau$ and the parameter $\delta$:

$$\begin{aligned}\tau^* &= \overline{ml}g\left(\sin\theta_1 + \delta\cos\theta_1\right) - I\left[\ddot{\theta}_1^{ref} - k_p\left(\theta_1 + \delta - \theta_1^{ref}\right) - k_v\left(\dot{\theta}_1 - \dot{\theta}_1^{ref}\right)\right] \\ &= \tau + Ik_p\delta + \overline{ml}g\delta\cos\theta_1\end{aligned} \tag{6.78}$$

The acceleration $\ddot{\theta}_1^*$ taking into account the disturbance can then be expressed as a function of the nominal acceleration $\ddot{\theta}_1$ without disturbance. To do this, we inject the modified control law (6.78) into the flywheel dynamics Eq. (6.67b). By identification, we substitute the value of the nominal system acceleration (6.70b):

$$\ddot{\theta}_2^* = \ddot{\theta}_2 + \frac{I + i_2}{I i_2}\left(I k_p \delta + \overline{ml} g \delta \cos\theta_1\right) \tag{6.79}$$

To obtain an estimate of the angular velocity of the inertia wheel at the end of the cycle as a function of the parameter $\delta$ Eq. (6.79) can be integrated over the cycle period:

$$\begin{aligned} \int_0^{t_c} \ddot{\theta}_2^*(t)dt &= \int_0^{t_c} \ddot{\theta}_2(t)dt + \int_0^{t_c} \frac{I+i_2}{I i_2}\left(I k_p \delta + \overline{ml} g \delta \cos\theta_1(t)\right) dt \\ \dot{\theta}_2^*(t_c) - \dot{\theta}_2^*(0) &= \dot{\theta}_2(t_c) - \dot{\theta}_2(0) + \frac{I+i_2}{i_2} k_p \delta t_c + \frac{I+i_2}{I i_2}\overline{ml} g \delta \int_0^{t_c} \cos\theta_1(t)dt \end{aligned} \tag{6.80}$$

In the nominal case, i.e., undisturbed case, given that the reference trajectory optimization parameter $p$ minimizes the inertia wheel speed at the end of the cycle, we can assume that $\theta_2(t_c) = 0$. Let us now consider the disturbed system. The estimation of the disturbance on the inertia wheel speed at the end of each cycle is used to calculate the control input $\tau^*$ (6.75) allowing the continuation of the reference trajectories during the next cycle. In steady state, we can therefore assume that the parameter $\delta$ models the effect of the disturbance on the system accurately enough to consider that the measured speed $\dot{\theta}_2^*(t_c)$ is close enough to the actual speed $\dot{\theta}_2(t_c)$. If we take these considerations into account in Eq. (6.80), we obtain:

$$\dot{\theta}_2^*(t_c) = \frac{I+i_2}{i_2} k_p \delta t_c + \frac{I+i_2}{I i_2}\overline{ml} g \delta \int_0^{t_c} \cos\theta_1(t)dt \tag{6.81}$$

To obtain the relationship between the flywheel speed $\dot{\theta}_2^*$ and the parameter $\delta$, the integral term in Eq. (6.81) containing $\theta_1$ must be expressed as a function of $\theta_1^*$. Indeed, the term $\cos\theta_1$ can be linearized as follows:

$$\cos\theta_1 \approx \cos\theta_1^* + \delta \sin\theta_1^* \tag{6.82}$$

Substituting (6.82) into (6.81) gives:

$$\dot{\theta}_2^*(t_c) = \frac{I+i_2}{i_2} k_p \delta t_c + \frac{I+i_2}{I i_2}\overline{ml} g \delta \int_0^{t_c} \left[\cos\theta_1^*(t) + \delta \sin\theta_1^*(t)\right] dt \tag{6.83}$$

After manipulation, Eq. (6.83) takes the following form, expanded with respect to $\delta$:

$$\delta^2 \frac{\overline{ml}g}{I} \int_0^{t_c} \sin\theta_1^*(t)dt + \delta\left(k_p t_c + \frac{\overline{ml}g}{I} \int_0^{t_c} \cos\theta_1^*(t)dt\right) - \frac{i_2}{I+i_2}\dot{\theta}_2^*(t_c) = 0 \tag{6.84}$$

Eq. (6.84) can be seen as a polynomial of degree 2 in $\delta$, namely:

$$a\delta^2 + b\delta + c = 0 \tag{6.85}$$

whose coefficients are:

$$\begin{aligned} a &= \frac{\overline{ml}g}{I} \int_0^{t_c} \sin\theta_1^*(t)dt \\ b &= k_p t_c + \frac{\overline{ml}g}{I} \int_0^{t_c} \cos\theta_1^*(t)dt \\ c &= -\dot{\theta}_2^*(t_c)\frac{i_2}{I+i_2} \end{aligned} \tag{6.86}$$

The roots of this polynomial can be calculated in real time at the end of each cycle. The first root corresponds to the variation in $\delta$ that compensates for the offset observed in the previous cycle, the offset estimation parameter is updated as follows:

$$\begin{cases} \delta_{k+1} = \delta_k + \Delta\delta_{k+1} \\ \delta_0 = 0 \end{cases} \tag{6.87}$$

The value $\Delta\delta_{k+1}$ is the root calculated at the end of the $k$th cycle, given the observed value of flywheel speed $\dot{\theta}_2^*(kt_c)$.

## 6.5 Numerical simulation results

Numerical simulations are carried out in a Matlab/Simulink® environment. The differential equations governing the system dynamics are integrated using the Runge–Kutta method (function ode45 from Matlab). The control law is calculated according to Eq. (6.68). The reference trajectory $\theta_1^{ref}(t)$ is described in Section 6.4.1. The time derivatives $\dot{\theta}_1^{ref}(t)$ and $\ddot{\theta}_1^{ref}(t)$ are obtained by differentiating $\theta_1^{ref}(t)$ with respect to time. Fig. 6.7 shows the simulator block diagram used to produce all the simulation results presented in this chapter. The optimal value of parameter $p$ is calculated using a nonlinear numerical optimization method (function fminbnd from Matlab). The search space is restricted to the interval $[p_{min}, p_{max}]$. External persistent disturbance is estimated using the method described in Section 6.3.2.

Several simulation scenarios are considered. In the first scenario, no external disturbance is introduced into the model apart from non-zero initial conditions.

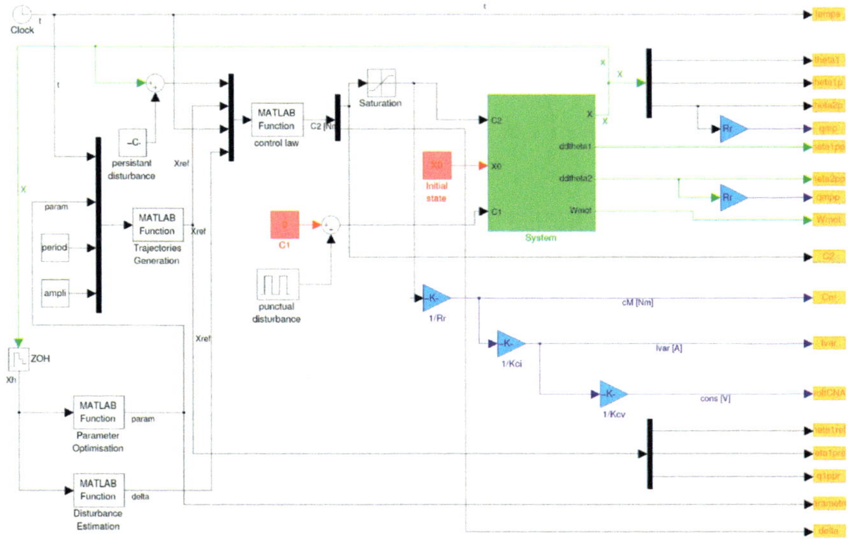


**FIGURE 6.7** View of Simulink block diagram used for numerical simulations.

**TABLE 6.2 Summary of the parameters of reference trajectories and feedback control.**

| Parameter | Description | Value | Unit |
|---|---|---|---|
| $A$ | amplitude of oscillations | 3 | deg |
| $\tau$ | period of oscillations | 2 | s |
| $k_p$ | proportional gain | 300 | $s^{-2}$ |
| $k_v$ | derivative gain | 20 | $s^{-1}$ |

In the second scenario, a point disturbance is introduced in the form of a disturbance torque $C_1 \neq 0$ at the base connection. The third scenario aims to highlight the effects of persistent disturbances, without taking them into account at the control level. Finally, the last scenario applies the proposed method for estimating and rejecting persistent disturbances.

### 6.5.1 Scenario 1: Nominal case

In this simulation, no external disturbance is considered. Table 6.2 summarizes the values used to generate the reference trajectories and the feedback gains of the proposed control law.

This simulation is initialized with the pendulum at rest, so the initial conditions are chosen accordingly:

- initial position of the pendulum: $\theta_1(0) = 10°$;
- initial pendulum speed: $\dot{\theta}_1(0) = 0\,\text{rd s}^{-1}$;

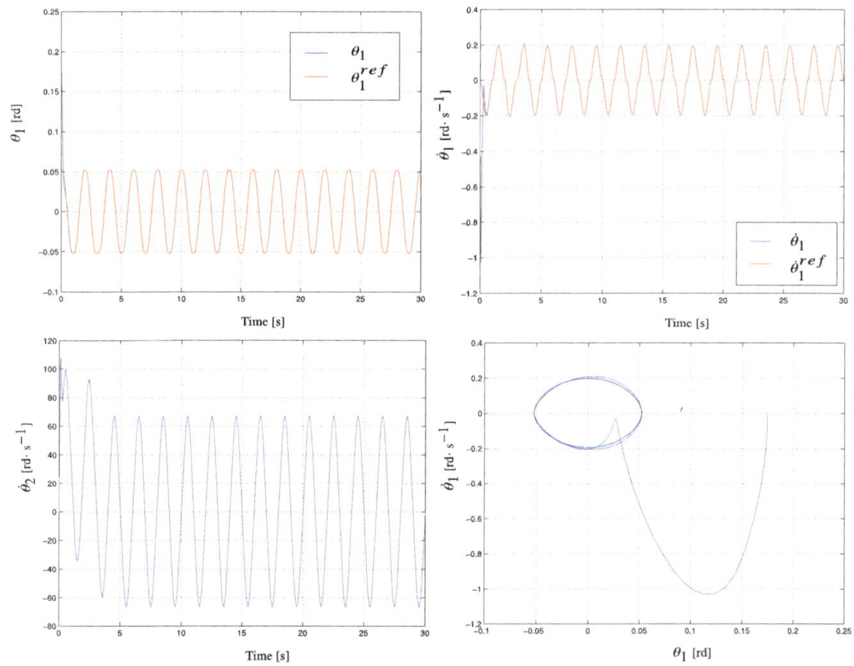


**FIGURE 6.8** Simulation results of scenario 1: Nominal case (positions, speeds, and phase portrait).

- initial flywheel speed: $\dot{\theta}_2(0) = 0\,\text{rd s}^{-1}$.

The choice of $\theta_1(0)$ is motivated by the physical structure of the robot. When the pendulum is at rest, the mechanical stop constrains the pendulum angle to approximately this value.

Figs. 6.8 and 6.9 show the obtained simulation results. Fig. 6.8 (top) shows the evolution of pendulum position and velocity as a function of time by solid lines, while the dotted lines represent the corresponding reference trajectories. This figure clearly shows the convergence of position and velocity towards their respective periodic reference trajectories. The flywheel speed versus time is also shown in Fig. 6.8 (lower left), where it can be seen that after two periods, the motor speed reaches a limit cycle oscillating around zero. The phase portrait $(\theta_1, \dot{\theta}_1)$ is illustrated in Fig. 6.8 (lower right), clearly showing convergence from the initial condition to a stable limit cycle.

Fig. 6.9 (top right) represents the evolution of the control input consisting of the voltage sent to the actuator motor (which is proportional to the torque generated by the motor). It is important to note that the control input remains well within the motor's permissible limits ($\pm$ 10 V). The evolution of the normalized parameter between $-1$ and 1, the result of the optimization, is shown in Fig. 6.9 (top left). This parameter converges to zero after two cycles. Fig. 6.9 (lower) shows the absolute motor speed as a function of the absolute motor torque and

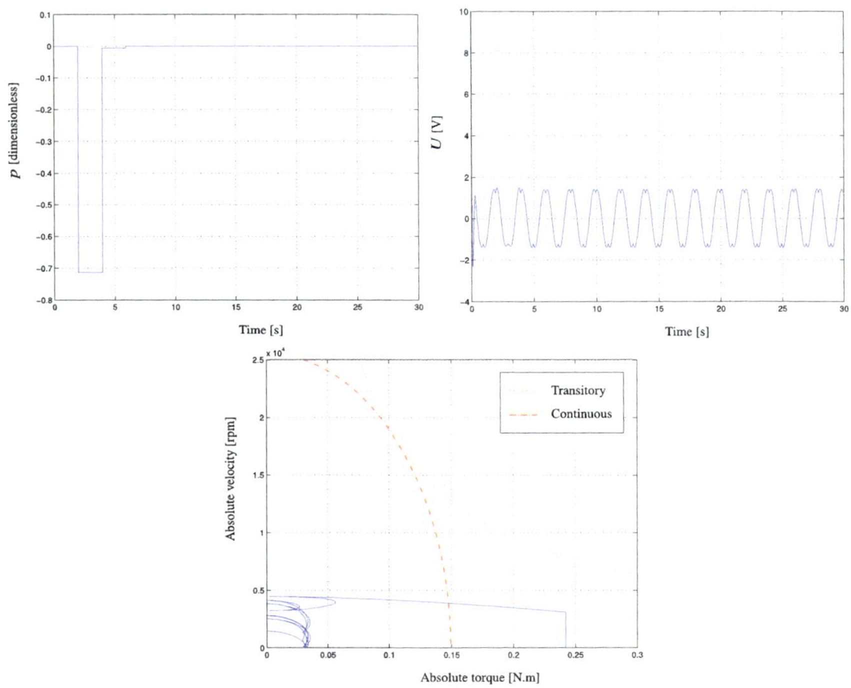


**FIGURE 6.9** Simulation results of scenario 1: Nominal case (optimization parameter, control input, and power admissibility).

the limits of the admissible region provided by the motor manufacturer. It can be seen that the trajectory remains within this admissible region.

## 6.5.2 Scenario 2: Point disturbance rejection

This simulation aims to show the effect of a point disturbance on the stability of the closed-loop system. The model and control law used are the same as for the previous scenario. The version of the controller used in this case is the one without persistent disturbance estimation. The disturbance is introduced in the form of a disturbance torque $C_1$, at time $t = 10$ s. The disturbance torque is a pulse with a duration of 0.35 s and an amplitude equal to 0.3 N. This impulse is sufficient to disturb the system. It occurs at a critical instant, at the end of the period, when the pendulum is at an extreme position.

The results of this simulation are shown in Figs. 6.10 and 6.11. Fig. 6.10 (top left) shows the evolution of the pendulum axis position as a function of time, as well as its reference trajectory. At $t = 10$ s we can see the disturbance that produces a deviation from the trajectory, which is quickly corrected by the controller. The angular velocity of the pendulum is shown in Fig. 6.10 (top right), where we also observe the effect of the perturbation under considera-

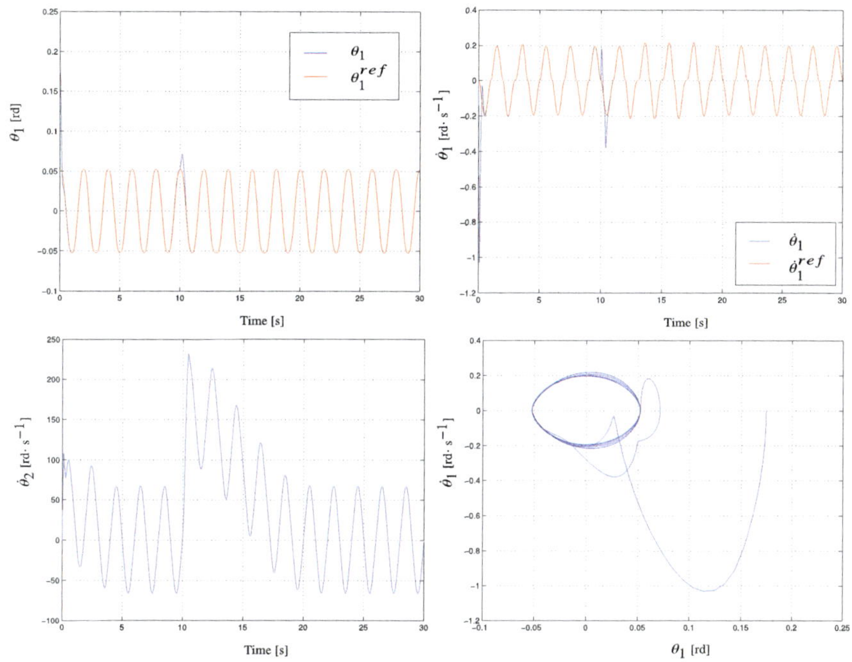


**FIGURE 6.10** Simulation results of scenario 2: Point disturbance rejection (positions, speeds, and phase portrait).

tion. Fig. 6.10 (lower left) shows the rotation speed of the flywheel. The effect of the disturbance is visible as a shift in the oscillations at the moment of disturbance. This is a consequence of the system's response, which corrects the tracking of the reference trajectories as best it can. The controller manages to bring the trajectories back around the limit cycle in just a few cycles. The pendulum's phase plane is illustrated in Fig. 6.10 (lower right), where it can be seen that the pendulum's trajectory joins a stable limit cycle, despite the disturbance under consideration.

Fig. 6.11 (top left) shows the evolution of the optimization parameter as a function of time. The controller's response can be seen as it corrects the reference trajectory. Although the parameter saturates at the start, the correction is sufficient to bring the steering wheel speed oscillations back to around zero. The control input is shown in Fig. 6.11 (top right) where we can observe the controller's immediate reaction to the disturbance, resulting in a spike in the trajectory. Finally, the motor power curve and the admissible region template are plotted in Fig. 6.11 (lower). It is worth noting that the trajectory remains within the admissible region, despite the disturbance. The controller is therefore able to reject a significant point disturbance, even if it occurs at a critical point in the system's evolution.

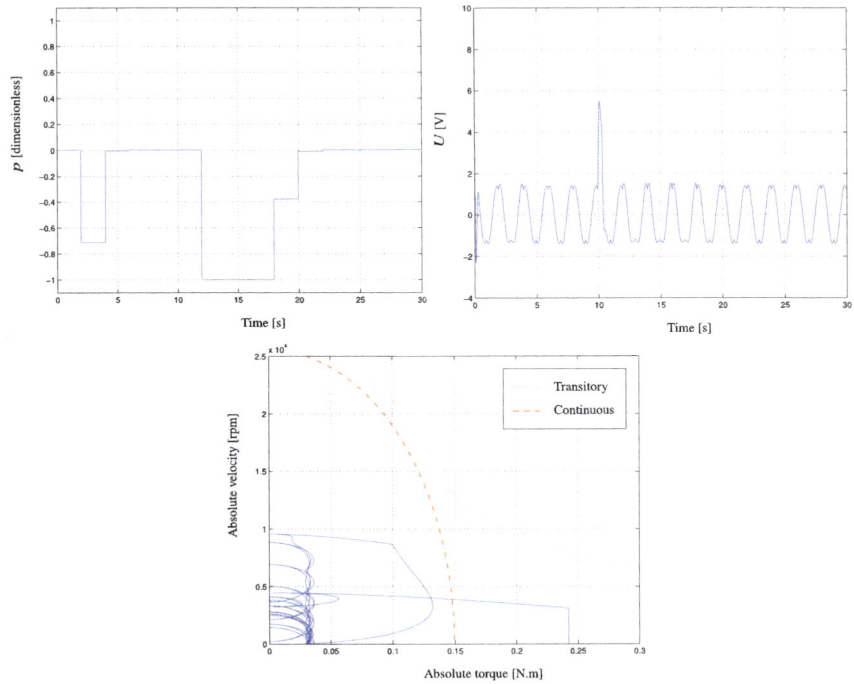


**FIGURE 6.11** Simulation results of scenario 2: Point disturbance rejection (optimization parameter, control input, and power admissibility).

## 6.5.3 Scenario 3: Persistent disturbances rejection

### *6.5.3.1 Persistent disturbances not estimated*

In this simulation, the system model and control law are the same as in the simulation of the nominal case without disturbance (Section 6.5.1). However, a persistent external disturbance is considered. This disturbance takes the form of a permanent offset of 0.02 rd in the angle measurement $\theta_1$ used in the control input calculation. This persistent disturbance is not taken into account in the synthesis of the control approach, in order to highlight the effect of a persistent disturbance on the closed-loop system and the need to take it into account in the control design through its estimation.

The results of this simulation scenario are shown in Figs. 6.12 and 6.13. The top curves of Fig. 6.12 show the evolution of the pendulum's position and velocity. The shift in the pendulum's position caused by the persistent disturbance is clearly visible. As this offset is not estimated, the controller has no way of compensating for this disturbance, resulting in a constant error in trajectory tracking. The flywheel speed versus time is shown in Fig. 6.12 (lower left). Note that the trajectory of the flywheel velocity diverges progressively. The effective trajectory $\theta_1(t)$ of the pendulum is not symmetrical with respect

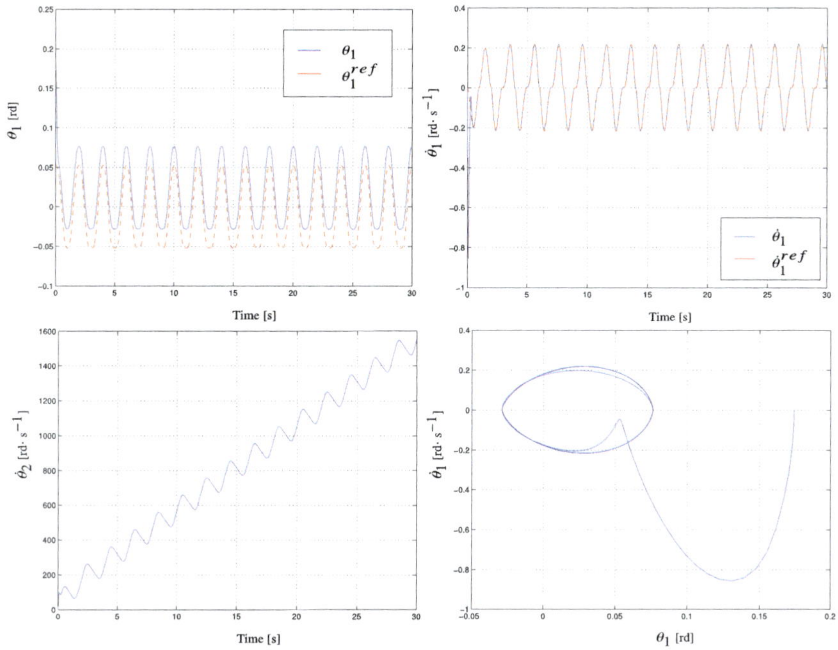


**FIGURE 6.12** Simulation results of scenario 3: Non-estimated persistent disturbance (positions, speeds, and phase portrait).

to the vertical due to measurement error, the trajectory of the torque applied by the motor to pursue this trajectory is not centered around zero. At the end of a cycle, the balance of forces developed by the motor is therefore not zero. As a result, the flywheel speed trajectory accumulates an offset with each cycle. The phase portrait $(\theta_1, \dot{\theta}_1)$ is plotted in Fig. 6.12 (lower right). Despite the persistent disturbance, we can clearly see convergence from the initial condition to a stable limit cycle of the pendulum, given that this part of the system is controllable. Fig. 6.13 (top right) shows the evolution of the control input as a function of time. The control input remains within the motor's permissible limits. The evolution of the normalized optimization parameter between $-1$ and 1 is shown in Fig. 6.13 (top left). It can be seen that this parameter is saturated from the second cycle onwards. Optimization of the reference trajectory parameter takes into account the steering wheel speed at the start of each cycle. This optimization assumes that a non-zero flywheel speed is due to point disturbances applied to the system. The optimal parameter chosen is therefore capable of bringing the flywheel speed at the end of the cycle back to around zero only if, during the next cycle, the system is no longer subject to this disturbance. Since the domain of the optimization parameter is bounded, saturation of the parameter is inevitable. The controller is therefore unable to compensate for such persistent disturbances. Fig. 6.13 (lower) shows the absolute motor speed as a function of

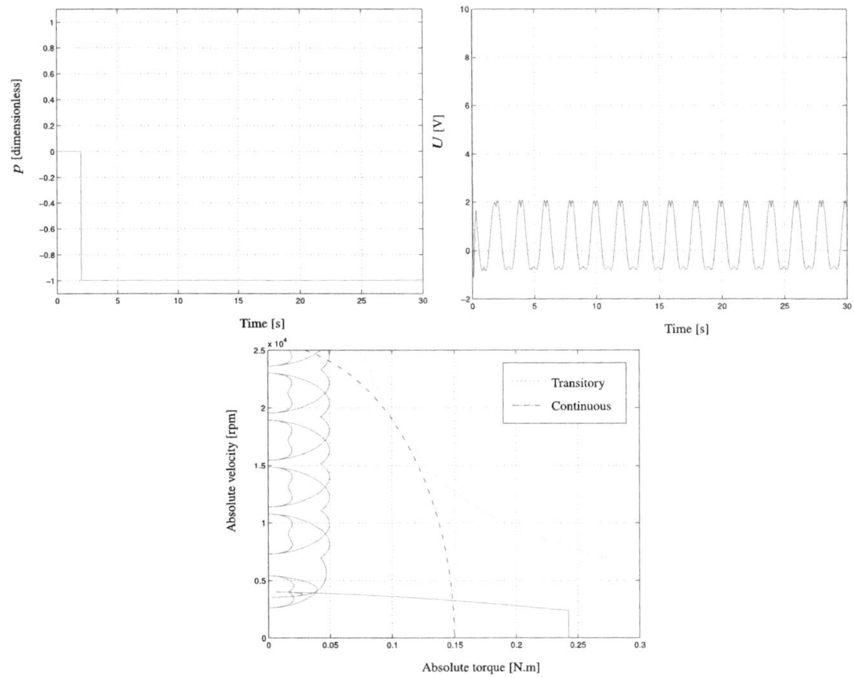


**FIGURE 6.13** Simulation results of scenario 3: Non-estimated persistent disturbance (optimization parameter, control input, and power admissibility).

the absolute motor torque and the permissible motor power region provided by the manufacturer. Although during simulation the trajectory remains within the admissible region, we note that the trajectory cycle progressively diverges. After a while, the motor will saturate and the pendulum's trajectory can no longer be tracked.

### 6.5.3.2 Estimated persistent disturbances

In this simulation, the model and controller parameters are the same as in the previous simulation. The same disturbance is applied in the same way. However, the controller based on persistent perturbation estimation described in Section 6.4.4 is used. The control law uses the estimated parameter $\delta$ to correct the measurement of the pendulum position $\theta_1$. Figs. 6.14 and 6.15 show the obtained results for this simulation. Figs. 6.14 (top left and right) show the evolution of the angular position and velocity of the pendulum. Although the offset caused by the persistent disturbance on the pendulum's position trajectory is present at the beginning of the simulation, we observe that this offset gradually decreases until it is completely compensated. The estimate of the offset becomes more accurate cycle after cycle, and the controller uses the estimate to correct this trajectory error. Fig. 6.14 (lower left) shows the evolution of the

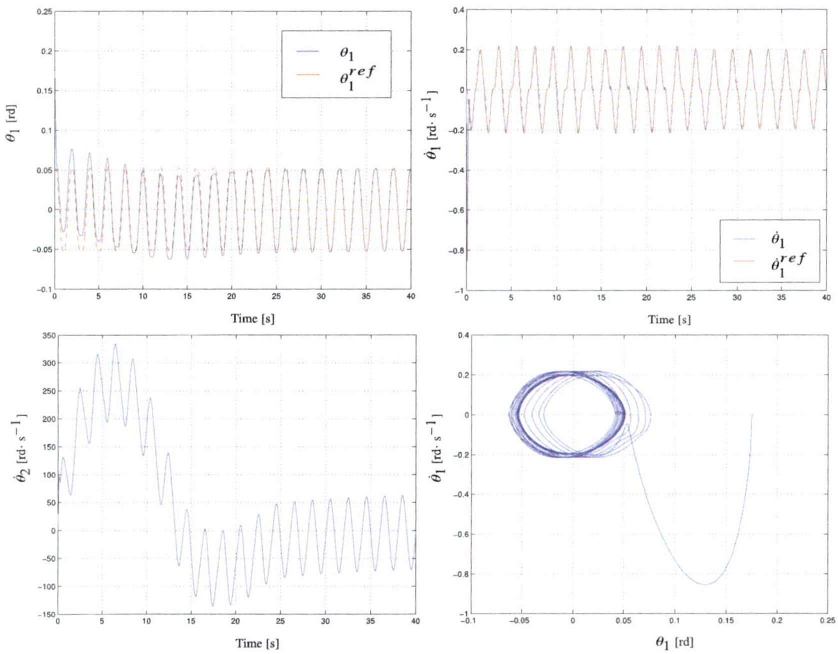


**FIGURE 6.14** Simulation results of scenario 4: Estimated persistent disturbance (positions, speeds, and phase portrait).

flywheel speed as a function of time. Note that the trajectory of the flywheel speed does not oscillate around zero at the beginning of the simulation. By estimating the disturbance, the controller is able to compensate for this and bring the flywheel speed trajectory back to zero within a few cycles. The phase plane $(\theta_1, \dot\theta_1)$ is shown in Fig. 6.14 (lower right). We can clearly see convergence from the initial condition to a stable limit cycle of the pendulum's trajectory. Fig. 6.15 (lower left) shows the evolution of the control input versus time. It is worth noting that the control signal remains within the admissible limits of the actuator. The evolution of the normalized parameter between $-1$ and 1, the result of optimization, is shown in Fig. 6.15 (top left). The optimization parameter is saturated at $-1$ at the start of the simulation and gradually approaches 0 as the disturbance estimate persists, whose evolution is shown in Fig. 6.15 (top right) becomes more precise. The estimated disturbance converges to the exact value of the disturbance in the model. Thanks to the estimated disturbance, the controller has a more accurate measure of the pendulum's position and is able to stabilize the system. Fig. 6.15 (lower right) shows the absolute motor speed as a function of the absolute motor torque, and the permissible motor region supplied by the manufacturer. The trajectory always remains within the admissible region.

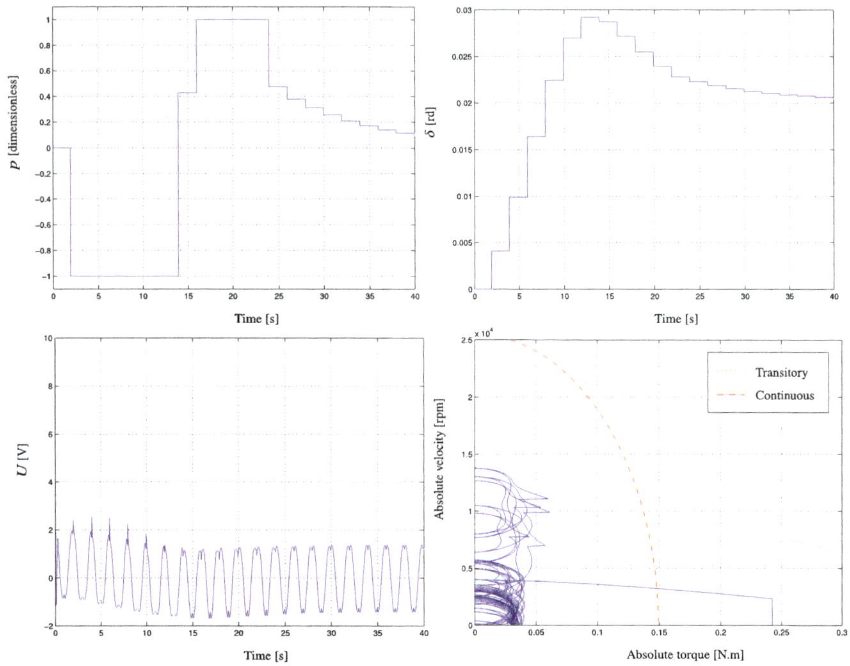


**FIGURE 6.15** Simulation results of scenario 4: Estimated persistent disturbance (optimization parameter, control input, and power admissibility).

## 6.6 Real-time experimental results

The proposed control scheme requires to resolve a real-time optimization problem at the start of each cycle. However, the platform cannot provide the computing time required to solve online this optimization problem. Optimization is therefore performed off-line for a grid of initial condition values for a cycle, and the results are stored in a lookup table. Given the system's $\sigma = (\theta_1, \dot{\theta}_1, \dot{\theta}_2)$ state at the start of a cycle, the controller linearly interpolates an optimization parameter value from the optimal values stored in the table for states adjacent to $\sigma$. The result is then obtained instantaneously, at the expense of optimization accuracy.

To validate the proposed control scheme, three experimental scenarios are considered. The first scenario shows the results of the control approach without external disturbance (nominal case). In the second, an external point disturbance is deliberately applied to the pendulum. In the last scenario, a persistent external disturbance is introduced into the system through a weight attached to the pendulum body.

### 6.6.1 Scenario 1: Nominal case without external disturbances

In this first scenario, no external disturbance is considered. However, the system is still subject to uncertainties and disturbances inherent in any real robotic

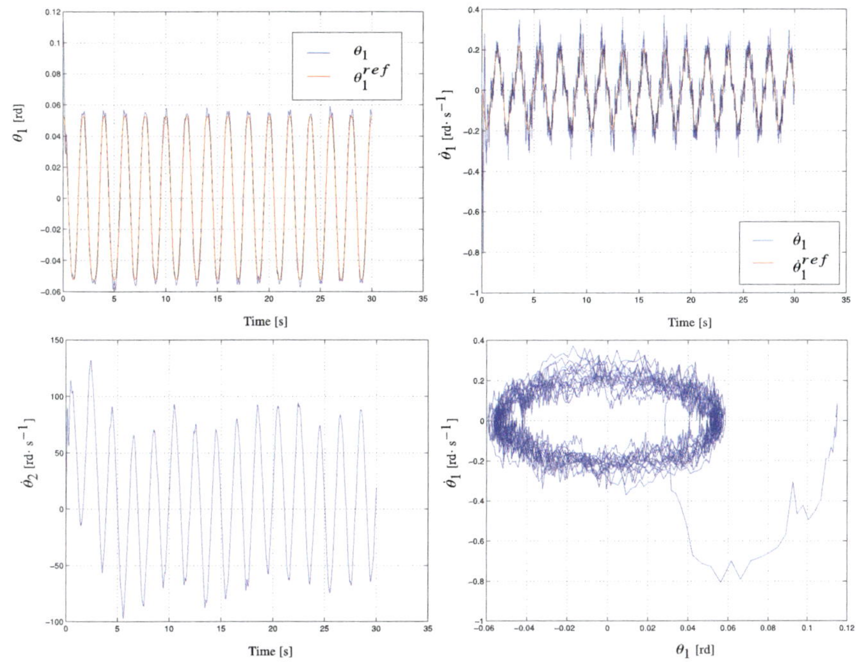


**FIGURE 6.16** Real-time experimental results scenario 1: Nominal case (positions, speeds, and phase portrait).

system: uncertainties about mechanical parameters, noisy measurements of the system state, etc. Figs. 6.16 and 6.17 show the results of this experiment. The position and velocity of the pendulum are shown in Figs. 6.16 (top left and right), respectively. The angular velocity of the pendulum $\dot{\theta}_1$ is very noisy, as it is calculated by numerical differentiation of the raw measurement of the pendulum angle, which is itself slightly noisy. Fig. 6.16 (lower left) shows the evolution of the motor speed versus time and also shows how optimizing the reference trajectories reduces the oscillation of the motor speed around zero in just a few cycles. The phase plane of the pendulum $(\theta_1, \dot{\theta}_1)$ is shown in Fig. 6.16 (lower right). The limit cycle is reached despite the presence of noise. The motor control input can be seen in Fig. 6.17 (top right), where we can see that it remains within the admissible range $\pm 10$ V. Fig. 6.17 (top left) shows the evolution of the optimization parameter over time. Finally, Fig. 6.17 (lower) shows the admissible power region of the actuator motor. From the evolution of the trajectories, it is possible to check that the motor power remains within the admissible region given by its manufacturer. This experiment starts with non-zero initial conditions. Despite this, the controller is able to converge the pendulum trajectories towards the reference trajectories, while maintaining the oscillations of the flywheel speed around zero.

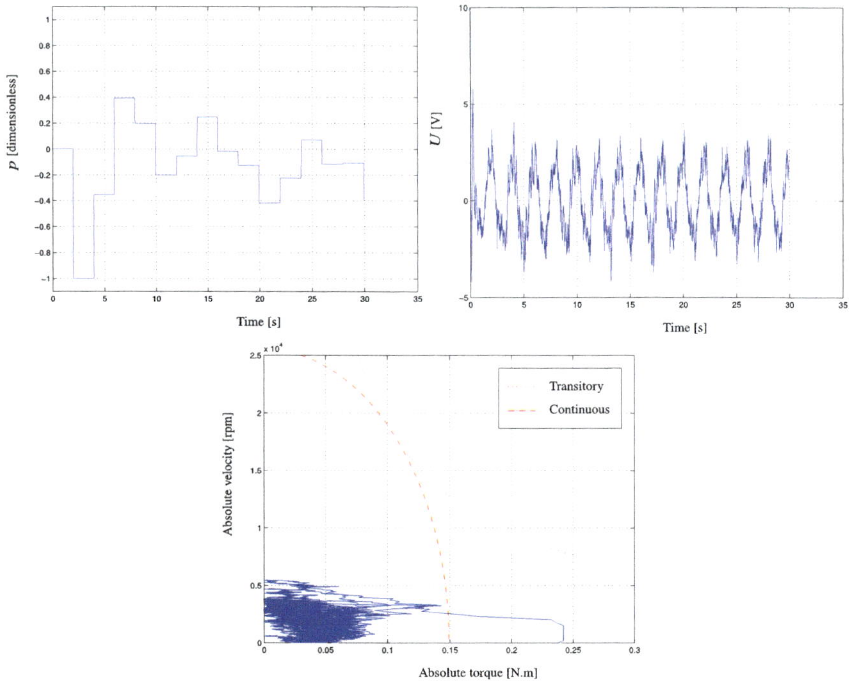


**FIGURE 6.17** Real-time experimental results of scenario 1: Nominal case (optimization parameter, control input, and power admissibility).

### 6.6.2 Scenario 2: Point external disturbance rejection

In this second scenario, external disturbances are introduced by pushing the pendulum body during the oscillating movement. These external forces generate punctual torques applied to the passive link between the pendulum and the frame. The disturbances occur approximately at time instants $t = 16$ s and $t = 33$ s.

Figs. 6.18 and 6.19 show the obtained experimental results for this scenario. In Fig. 6.18 (top left), the angular position of the pendulum as well as its reference trajectory are depicted. The angular velocity of the pendulum and its reference trajectory are displayed in Fig. 6.18 (top right). The effect of external point disturbances can be seen on these curves in the form of peaks appearing at the instants of application of the disturbing torques. The compensation of external disturbances can be seen in Fig. 6.18 as well as in Fig. 6.19 (lower left), which shows the evolution of the flywheel speed over time.

Perturbations cause a shift in the oscillations of the flywheel velocity trajectory. The effect of optimizing the reference trajectory is also clearly visible, bringing the oscillation back to zero within a few cycles.

The pendulum's phase portrait is depicted in Fig. 6.18 (lower right). The trajectory joins the limit cycle after each applied disturbance, demonstrating the

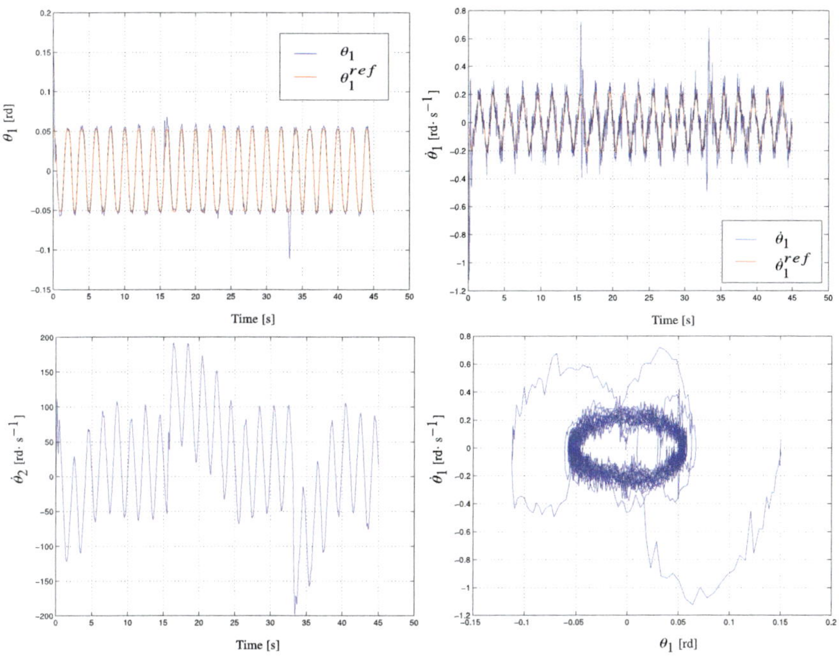


**FIGURE 6.18** Real-time experimental results scenario 2: Point external disturbance rejection (positions, speeds, and phase portrait).

ability of the proposed controller in terms of disturbance rejection. Fig. 6.19 (top right) displays the evolution of the motor control voltage. We can observe the controller's immediate reactions toward external disturbances, rapidly returning the pendulum to its reference trajectory. Fig. 6.19 (top left) depicts the evolution of the optimization parameter versus time. Here, the controller's reaction is slower, since the optimization parameter is updated only once per cycle. Finally, the motor power admissibility is represented by the evolution of rotation speed as a function of motor torque in Fig. 6.19 (lower) and the area of the admissible region.

The controller is thus able to maintain the system around the reference trajectories and reject external disturbances introduced punctually as torques applied to the pendulum body.

### 6.6.3 Scenario 3: Persistent external disturbances rejection

This scenario aims to demonstrate the effectiveness of the algorithm for estimating and rejecting constant and persistent disturbances (presented in Section 6.3.2). The controller parameters are the same as in the two previous scenarios, but the persistent disturbance is estimated and integrated with the pendulum angle measurement when calculating the control law. The estimation

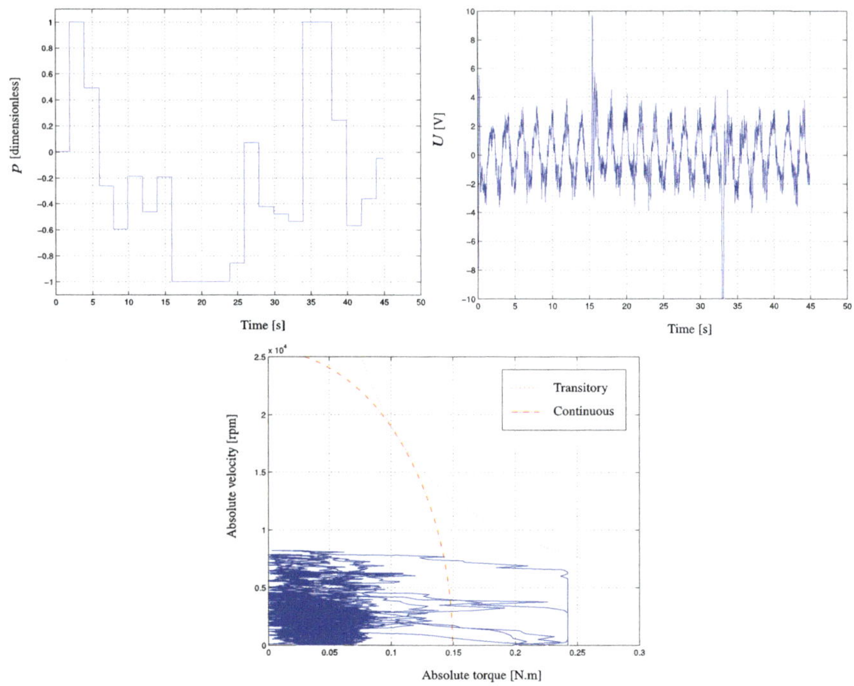


**FIGURE 6.19** Real-time experimental results of scenario 2: Point external disturbance rejection (optimization parameter, control input, and power admissibility).

of persistent disturbances requires the calculation of two integrals over a complete cycle in Eq. (6.84). These integrals are calculated numerically by summing successive values over the cycle.

The persistent disturbance is introduced whose nature is not taken into account by the model. An object (a clamp) of unknown weight is attached to the pendulum body on one side of the vertical axis, then removed at approximately $t = 30$ s. The weight fixed asymmetrically on the pendulum body induces a disturbing external torque acting on the passive joint between the pendulum and the frame. Although the intensity of this torque varies over time due to the movement of the pendulum axis, its effect on the system over a complete period is constant, since the pendulum's trajectory is periodic and symmetrical. The persistent disturbance thus introduced is constant from the point of view of the estimator, since it is observed only at the end of the cycle.

The results of this experiment are shown in Figs. 6.20 and 6.21. The evolution of the pendulum's angular position and velocity versus time are shown in Figs. 6.20 (top left and right) with their reference trajectories. From these curves, it is possible to observe the convergence towards a stable limit cycle corresponding to the periodic reference trajectories. We can also observe the offset introduced by the controller on the pendulum position trajectory to com-

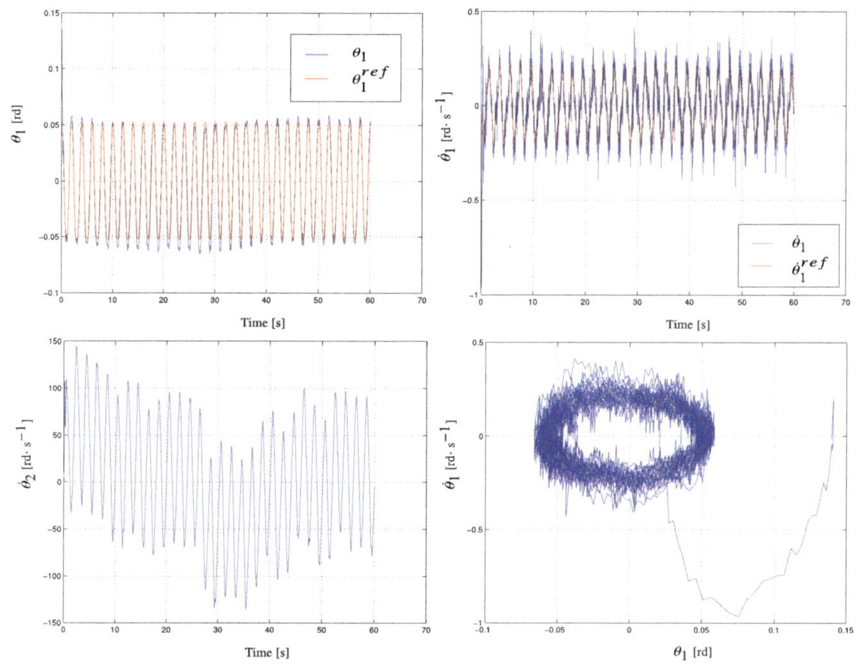


**FIGURE 6.20** Real-time experimental results scenario 3: Persistent external disturbance rejection (positions, speeds, and phase portrait).

pensate for the effect of the weight added to the pendulum body in the first half of the experiment. The angular velocity of the flywheel is shown in Fig. 6.20 (bottom left), where we can observe the effects of the weight at the beginning of the experiment. Just after $t = 30$ s we can see the effect of compensation for this disturbance. The controller adapts within a few cycles, and after each disturbance brings the oscillations of the flywheel speed back to zero, guaranteeing the stability of the limit cycle. The convergence of the pendulum towards the stable limit cycle can be seen on the phase plane shown in Fig. 6.20 (lower right).

Fig. 6.21 (top left) shows the evolution of the trajectory optimization parameter over time. The saturation of the parameter is observed after each perturbation (addition and removal of the perturbation weight), and how persistent perturbation estimation keeps the parameter within its domain after a few cycles. The evolution of the perturbation estimation is shown in Fig. 6.21 (top right). The value towards which this estimate converges just before the weight is removed is a virtual shift of the pendulum that corresponds exactly, in its effects on system dynamics, to the weight added to the pendulum. The estimate quickly converges to (approximately) zero after the weight is removed. Fig. 6.21 (bottom left) shows the evolution of the motor control voltage. Finally, the ad-

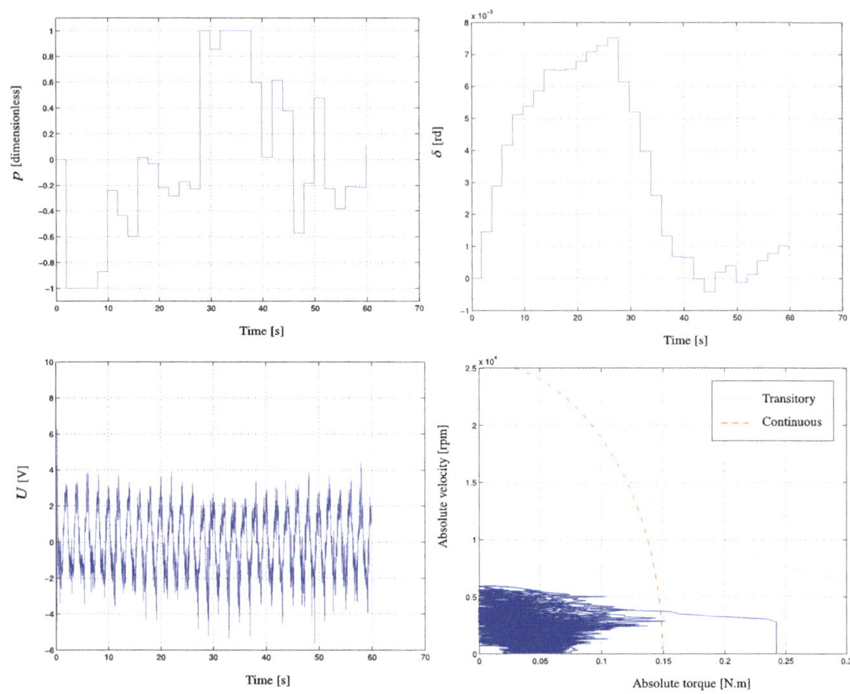


**FIGURE 6.21** Real-time experimental results of scenario 3: Persistent external disturbance rejection (optimization parameter, control input, and power admissibility).

missibility of the motor power is verified in Fig. 6.21 (lower right), where it should be noted that the trajectories remain within the admissible region.

According to the obtained results, the controller can maintain system stability even when the disturbance is different from the one considered in the modeling. The types of disturbance that the controller can reject extend to variable (periodic) disturbances, whose effect on system dynamics over a complete cycle is constant.

## 6.7 Conclusion

In this chapter, the first solution for generating stable limit cycles on an underactuated mechanical system has been presented in the general case. The principle of partial linearization (used in this method) was explained in detail and applied to the case of a particular underactuated mechanical system, the inverted double pendulum, named in two variants depending on the location of the system's actuator: the acrobot and the pendubot. Analysis of these two mechanical systems has shown that partial feedback linearization alone fails to stabilize the entire system. The optimization was then described with the aim of stabilizing the system's internal dynamics the system's internal dynamics by bringing the system as close as possible to a limit cycle. Finally, the model was modified to take account of persistent disturbances that are not fully rejected with a nominal model. The resulting model effectively rejects both persistent and punctual disturbances.

## References

[1] L. Ovalle, H. Ríos, M. Llama, L. Fridman, Continuous sliding-mode output-feedback control for stabilization of a class of underactuated systems, IEEE Transactions on Automatic Control 67 (2022) 986–992.

[2] B. Salamat, D. Bencic, G. Elsbacher, C. Seidel, A. Tonello, Investigating integral reinforcement learning to achieve asymptotic stability in underactuated mechanical systems, IEEE Robotics and Automation Letters 9 (2024) 191–198.

[3] M. Zhai, N. Sun, T. Yang, Y. Fang, Underactuated mechanical systems with both actuator and actuated/unactuated state constraints: a predictive control-based approach, IEEE/ASME Transactions on Mechatronics 28 (2023) 1359–1371.

[4] J. Romero, I. Gandarilla, V. Santibanez, B. Yi, A constructive procedure for orbital stabilization of a class of underactuated mechanical systems, IEEE Transactions on Control Systems Technology 30 (2022) 2698–2706.

[5] E. Franco, P. Arpenti, A. Donaire, F. Ruggiero, Integral IDA-PBC for underactuated mechanical systems subject to matched and unmatched disturbances, IEEE Control Systems Letters 8 (2024) 568–573.

[6] S. Andary, A. Chemori, S. Krut, Estimation-based disturbance rejection in control for limit cycle generation on inertia wheel inverted pendulum testbed, in: IROS'09: Proceedings of the 2009 IEEE/RSJ International Conference on Intelligent Robots and Systems, 2009, pp. 1302–1307.

[7] S. Andary, A. Chemori, S. Krut, Control of the underactuated inertia wheel inverted pendulum for stable limit cycle generation, RSJ Advanced Robotics 23 (2009) 1999–2014.

[8] S. Andary, A. Chemori, S. Krut, Stable limit cycle generation for underactuated mechanical systems, application: inertia wheel inverted pendulum, in: IROS'08: IEEE/RSJ International Conference on Intelligent RObotsand Systems, 2008.

[9] M. Chowdhury, G. Shafiullah, SSR mitigation of series-compensated DFIG wind farms by a nonlinear damping controller using partial feedback linearization, IEEE Transactions on Power Systems 33 (2018) 2528–2538.

[10] A. Accetta, F. Alonge, M. Cirrincione, F. D'Ippolito, M. Pucci, R. Rabbeni, A. Sferlazza, Robust control for high performance induction motor drives based on partial state-feedback linearization, IEEE Transactions on Industry Applications 55 (2019) 490–503.

[11] T. Roy, M. Mahmud, W. Shen, A. Oo, An adaptive partial feedback linearizing control scheme: an application to a single machine infinite bus system, IEEE Transactions on Circuits and Systems. II, Express Briefs 67 (2020) 2557–2561.

[12] R. Marino, P. Tomei, A class of globally output feedback stabilizable nonlinear nonminimum phase systems, IEEE Transactions on Automatic Control 50 (2005) 2097–2101.

[13] X. Hu, Y. Zhao, B. Xu, C. Hu, Robust adaptive fuzzy tracking control for uncertain MIMO nonlinear nonminimum phase system, IEEE Transactions On Systems, Man, And Cybernetics: Systems 50 (2020) 2017–2028.

[14] S. Nazrulla, H. Khalil, Robust stabilization of non-minimum phase nonlinear systems using extended high-gain observers, IEEE Transactions on Automatic Control 56 (2011) 802–813.

[15] M. Benosman, F. Liao, K. Lum, J. Wang, Nonlinear control allocation for non-minimum phase systems, IEEE Transactions on Control Systems Technology 17 (2009) 394–404.

[16] L. Sciavicco, B. Siciliano, Modeling and Control of Robot Manipulators, McGraw Hill, 1996.

[17] M. Spong, Underactuated mechanical systems, in: Control Problems in Robotics and Automation, 1998.

[18] M. Spong, Partial feedback linearization of underactuated mechanical systems, in: Proceedings of the IEEE/RSJ/GI International Conference on Intelligent Robotsand Systems '94, 1994.
[19] M. Spong, The control of underactuated mechanical systems, in: Proceedings of the First International Conference on Mecatronics, 1994.
[20] H. Khalil, Nonlinear Systems, second edition, Prentice Hall, Upper Saddle River, 1996.
[21] Y. Lu, Z. Du, Y. Li, Y. Zhang, Nonlinear damping property and limit cycle of VSC, IEEE Transactions on Power Delivery 37 (2022) 4485–4488.
[22] E. Moulay, Rational Lyapunov functions and stable algebraic limit cycles, IEEE Transactions on Automatic Control 59 (2014) 1077–1081.
[23] S. Kumar, O. Makarenkov, R. Gregg, N. Gans, Stability of time-invariant extremum seeking control for limit cycle minimization, IEEE Transactions on Automatic Control 67 (2022) 5017–5024.
[24] G. Garofalo, C. Ott, Energy-based limit cycle control of elastically actuated robots, IEEE Transactions on Automatic Control 62 (2017) 2490–2497.
[25] G. Garofalo, C. Ott, Limit cycle control using energy function regulation with friction compensation, IEEE Robotics and Automation Letters 1 (2016) 90–97.
[26] X. Lou, Y. Li, R. Sanfelice, Notions, stability, existence, and robustness of limit cycles in hybrid dynamical systems, IEEE Transactions on Automatic Control 69 (2024) 4910–4925.
[27] B. Siciliano, L. Sciavicco, L. Villani, G. Oriolo, Robotics: Modelling, Planning and Control, 2009.
[28] B. Lu, Y. Fang, Online trajectory planning control for a class of underactuated mechanical systems, IEEE Transactions on Automatic Control 69 (2024) 442–448.

Chapter 7

# Nonlinear model predictive control

## Contents



## 7.1 Introduction

Scientists attempt to resolve many problems of our civilization in piercing the secrecy of living creatures' movements. With minimum matter and energy, animals obtain a maximal effect. Constantly fleeing or hunting, animals have developed complex movements that allow them to move with noteworthy effectiveness. A kangaroo's jump is particularly well adapted to irregular ground, providing that they are sufficiently free to allow vertical movements. Kangaroos are able to realize real physical prowess. They can jump up to thirteen meters long and three meters in height! When they are fleeing, their displacement speed can reach 80 km/h. Moreover, the energetic assessment of this means of displacement is extremely good. The study of a kangaroo's oxygen consumption has shown that jumping is one of the most economic means of displacement of

Control of Underactuated Mechanical Systems. https://doi.org/10.1016/B978-0-44-324020-1.00016-6

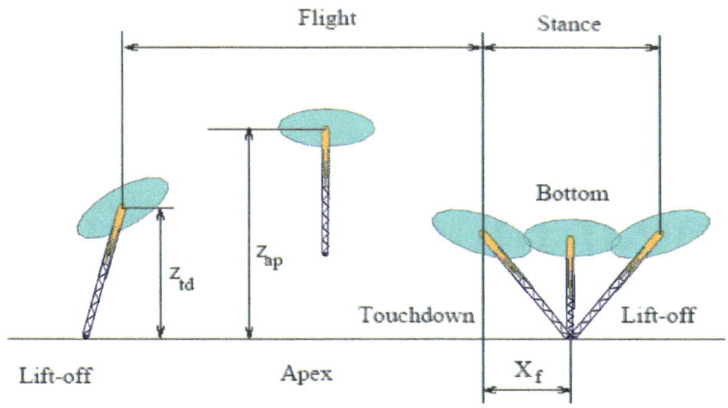


**FIGURE 7.1** Description of different phases of the jump cycle.

the animal kingdom. At the touchdown, Achilles tendons store up to 70% of the jump energy to release it in the next jump. Jumping requires a physical exertion 40% less than running.

## 7.2 Kangaroo underactuated hopping robot

### 7.2.1 Jump cycle description

Hopping robots are variable structure hybrid underactuated systems [22]. Control of a jump requires a dynamic modeling [1] of the robot for the different phases of the jump cycle (cf. Fig. 7.1). The number of degrees of freedom of the system changes from one phase to another.

The jump cycle can be decomposed into two main phases and two transitions, as illustrated in Fig. 7.1:

**Lift-off transition:** This characterizes the switch between the stance phase and the flight phase. At that moment, the impulse of the leg during the contact phase gives enough energy to throw the robot up in the flight phase.

**Flight phase:** A main phase in which the robot is in the air and describes a ballistic movement. It is characterized by the instant when the robot reaches the maximum height, called the Apex.

**Touchdown transition:** This is characterized by the switch between the flight phase and the stance phase.

**Stance phase:** A main phase in which the robot is in contact with the ground and follows a swing motion. The moment marked by a vertical position of the leg is called Bottom.

### 7.2.2 Kangaroo hopping robot design

Consider a Kangaroo robot illustrated in Fig. 7.2. It is a one-legged planar hopping robot. Its mechanical structure (cf. Fig. 7.2) has the following five degrees of freedom:

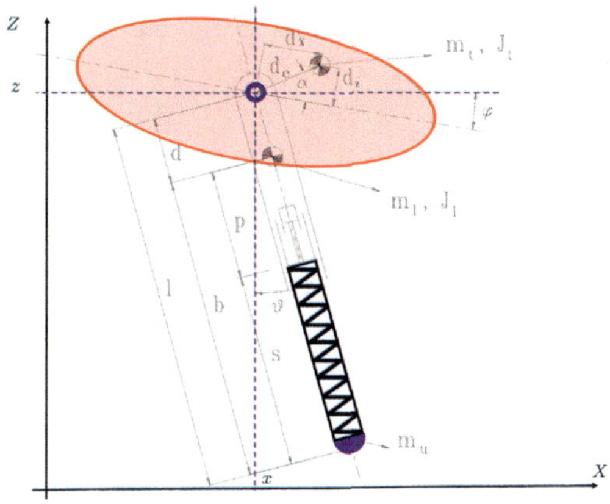

**FIGURE 7.2** View of Kangaroo robot's mechanical structure and its dynamic parameters [5].

$x, z$: two degrees of freedom for the Cartesian position of the body in the $XOZ$ plane (also called the sagittal plane);
$l$: one degree of freedom for the length of the leg;
$\phi$: one degree of freedom for the angle of the body in the sagittal plane;
$\nu$: one degree of freedom for the angle of the leg in the sagittal plane.

The system includes three bodies, they are considered as three point masses located at their respective center of gravity: $m_t$ for the body of the robot, $m_l$ for the leg, and $m_u$ for the foot. The center of gravity of the body $m_t$ is located at a distance $d_e$ from the hip's pivot and forms an angle $\alpha$ with the longitudinal axis of the robot's body (cf. Fig. 7.2). The center of mass of the leg $m_l$ is located at a distance $d$ from the hip of the robot. The first extremity of the leg spring is fixed to the upper part of the leg, at a distance $p$ from the center of gravity of the leg. The other one is fixed at the foot of the robot, the tube containing the spring slides freely in the upper part of the leg. The spring is characterized by a stiffness coefficient $kl$ and a length $s$. The length $s_0$ represents the spring length at rest. Finally, the length $l$ represents the total length of the leg ($l_0$ at rest) and defines the distance from the center of gravity of the foot $m_u$ to the hip. The moments of inertia of the body and the leg are denoted, respectively, by $J_l$ and $J_t$.

The robot is equipped with two actuators:

- **an angular actuator** at the hip that acts on the body's orientation relative to the leg;
- **a linear actuator** in the leg made of a compliant electrical actuator that acts on elongation and contraction of the leg.

The linear actuator in the robot's leg is associated with a spring mounted in parallel that stores energy from the jump. It can be compared to the biological Achilles tendon of the kangaroo animal. The actuator exerts a force between

**TABLE 7.1** Summary of passive model parameters of Kangaroo robot (COG: center of gravity).

| Symbol | Description | Value |
|---|---|---|
| $m_t$ | Robot's body mass | 12.5 kg |
| $m_l$ | Robot's leg mass | 3 kg |
| $m_u$ | Mass of robot foot | 0.5 kg |
| $J_t$ | Moment of inertia of the robot's body | 1 kg/m$^2$ |
| $J_l$ | Moment of inertia of the robot's leg | 0.2 kg/m$^2$ |
| $d_e$ | Distance from hip to COG of robot body | 0.1 m |
| $\alpha$ | Angle between the robot body and the COG of the robot body | 0 rad |
| $d$ | Distance between the hip and the COG of the robot leg | 0.2 m |
| $p$ | Distance from COG of the robot leg to the spring attachment point | 0.25 m |
| $\dot{p}$ | Displacement speed of the spring attachment point w.r.t. the hip | 0 m/s |
| $s_0$ | Length of spring at rest | 1 m |
| $g$ | Gravitational acceleration | 9.81 m/s$^2$ |
| $k_l$ | Spring stiffness constant | 5.05 kN/m |
| $c_l$ | Leg friction constant | 18 |
| $c_h$ | Hip friction constant | 6 |

the upper and lower leg parts. Since the number of degrees of freedom of the robot exceeds the number of control inputs, the system is underactuated. The modeled viscous frictions are located, respectively, at the hip (characterized by the coefficient $c_h$) when the body moves relative to the leg, and at the level of the leg (characterized by the coefficient $c_l$) where the lower part slides in the upper one.

The geometric and dynamic parameters of the robot are summarized in Table 7.1.

### 7.2.3 Robot's dynamic modeling

The Kangaroo robot has five degrees of freedom (cf. Section 7.2.2), the vector of generalized coordinates that enables us to describe the system is given by: $q = \begin{bmatrix} q_1 & q_2 & q_3 & q_4 & q_5 \end{bmatrix}^T = \begin{bmatrix} x & z & l & \phi & \upsilon \end{bmatrix}^T$. The robot is modeled on the two main phases of the jump cycle.

#### *7.2.3.1 Flight phase*

During the flight phase, the robot is in a ballistic movement and is not subject to any constraint. The Lagrangian dynamic model [1,23] of the robot is written in the following matrix form:

$$M(q)\ddot{q} + N(q,\dot{q})\dot{q} + F(\dot{q}) + G(q) = Su \tag{7.1}$$

where $M$ is the inertia matrix, $N$ is the matrix of Coriolis and centrifugal terms, $F$ is the friction vector, $G$ is the gravity vector, $S$ is a matrix of dimensions $5 \times 2$ that distributes the control inputs on the generalized coordinates, $u$ is the two-dimensional input control vector corresponding to the force exerted by the linear actuator located at the leg and the torque exerted by the angular actuator located at the hip.

#### 7.2.3.2 Stance phase

The dynamic model during the stance phase is obtained from the model of the flight phase by adding constraints of contact with the ground [1]. These constraints introduce an additional term representing the forces of contact with the ground. Assuming that friction between the foot and the ground is sufficient to prevent any sliding, the constraints of ground contact can be expressed as follows:

$$\begin{cases} x_f = q_1 + q_3 \sin(q_5) = cte \\ z_f = q_2 - q_3 \cos(q_5) = 0 \end{cases} \tag{7.2}$$

where $(x_f, z_f)$ are the Cartesian coordinates of the robot's foot.

These constraints are derived twice and integrated with the dynamics (7.1). The resulting dynamic model of the stance phase is then given by:

$$\begin{cases} M(q)\ddot{q} + N(q,\dot{q})\dot{q} + F(q) + G(q) = U + J(q)^T\lambda \\ J(q)\ddot{q} + P(q,\dot{q}) = 0 \end{cases} \tag{7.3}$$

where $J(q)$ is the Jacobian matrix of contact constraints with the ground, $P(q,\dot{q}) = \dot{J}(q,\dot{q})\dot{q}$ and $\lambda$ is the vector of Lagrange multipliers related to the contact forces.

#### 7.2.3.3 Transitions between main phases

In the previous section, the Kangaroo robot is modeled in its two main phases, namely the flight phase and the stance phase. (See Fig. 7.3.)

During the stance phase, the vertical component of the contact force exerted on the ground $\lambda_2$ can only be positive. We are in the case of a unilateral constraint, that is to maintain contact with the ground, the hopping robot has to exert only a positive force at the contact and the ground does not retain the robot's foot if the latter lifts off. The instant of transition from one phase to another is detected by analyzing the vertical component of the force generated by the constraints. When $\lambda_2 \leq 0$, the robot lifts off and enters the flight phase.

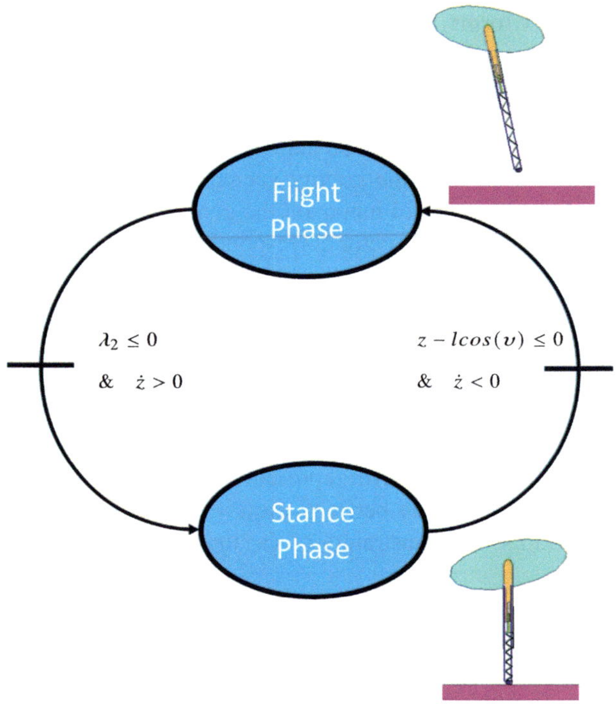


**FIGURE 7.3** Transitions between main phases.

The transition from flying to the contact phase is validated when the robot's foot touches the ground, which is $z - lcos(\nu) \leq 0$. Other conditions on velocity are added to overcome possible numerical problems (e.g., chattering). The robot must be, respectively, in the ascending phase to lift-off and in the descendant phase to touchdown. This prevents inadvertent crossing between the main phases.

## 7.3 Control problem and related works

Hopping robots [28] have been widely studied during the last decades. Usually made of a body and a compliant leg, these systems are capable of moving in one, two or three dimensions [2], alternating phases of ground contact and flight. Consequently, they can be regarded as *variable structure systems* [3], so that mechanical constraints can vary as the contact condition changes. The displacement of one-legged hopping robots has been obtained by the design of two types of machines: passive robots and robots equipped with controlled actuators.

There are some studies that were conducted on passive movement of planar hopping robots. These are robots that have a compliance at the leg and hip. The choice of adequate initial conditions and mechanical parameters enable these passive systems to move on a slope under the effect of gravity [4]. However, their mobility is reduced because of friction that provokes energy loss in the system [5]. By adding actuators, some controls were specifically designed to compensate the effects of friction [6,7].

One of the pioneers of hopping robots control is M. Raibert. He was at the origin of most current controls in the field of the hopping robots. In the early 1980s, he imagined a command to control the movement of a planar one-legged hopping robot [2]. In his control system, the thrust for jumping, the forward speed, and the body attitude are treated as three separate control problems.

Indeed, during the last two decades, different control approaches have been proposed to deal with these three issues. The control of the thrust of the leg [8] has been the subject of study of vertical hopping robots. At first, open-loop control was developed to achieve a constant hopping height [9]. Then, research focused on closed-loop control. In [3], the authors show that the height of a jump converges to a stable limit cycle by regulating the total energy (Hamiltonian) of the system. On the other hand, Harbick et al. have determined the vertical velocity of the robot body needed at the take-off instant to reach a specified apex height [10], then they have expanded their study to control the jump height for planar displacement [11]. The forward speed control of a hopping robot [25–27] has been particularly studied. To deal with this problem, a SLIP (Spring Loaded Inverted Pendulum) model with a point body mass and a leg is generally used. The proposed approaches consist in computing of the position of the foot at touchdown that enables to accelerate or to slow down the robot during the following stance phase [2]. In [12], an analytical study of the stability of the forward velocity equilibrium behavior has been conducted. It allows us to formulate a new law to control the foot position at touchdown. In [13], a polynomial expression of the angular leg trajectory is used as a reference for an inverse dynamic controller during the flight phase. Inverse dynamic control has also been adopted in [7] for forward speed control as well as for body attitude control during the stance phase [7,13]. Sato and his coworkers have proposed a new control approach using a SLIP model composed of a single actuator at the hip's level. The proposed control approach allows generation of stable periodic motions in simulation [14] as well as in real-time experiments [15].

The previously listed control approaches assume that the ground is continuous and flat. However, it is generally not the case when the robot moves in a real environment where the ground can be irregular. Therefore trajectory planning will be necessary to improve the efficiency of the control approach. Research in this direction uses generally a SLIP model to simplify the system dynamics. In [16], a sequence planner is proposed, which aims to find sequences of foot placements on a flat artificial ground with holes. Recently, reactive planning techniques have been proposed [17], which are based on a characterization of the attractive and feasible goal sets of potential footholds on a piecewise flat surface map with holes.

Other recent research works have been particularly focused on the biomimetism [25–27]. The physiological characteristics of animals have directly been imitated in insect hopping robots [19] or mammals [18]. On the one hand, this allows us to significantly advance the terrestrial techniques of displacement [20]. On the other, some of these studies will certainly improve knowledge about the physiology of living beings, including humans [21].

## 7.4 Background on model predictive control

Nonlinear Model Predictive Control (NMPC) is an advanced control strategy used in various engineering fields, particularly in process control, robotics, and aerospace. It builds on the principles of Model Predictive Control (MPC), which is a method that utilizes a model of the system to predict the future behavior and optimize control actions.

### 7.4.1 Key elements of NMPC

**Modeling:** NMPC relies on a dynamic model of the system, often represented by nonlinear differential or difference equations. This model captures the essential dynamics and behaviors of the system being controlled.

**Optimization:** At each time step, NMPC solves an optimization problem over a finite prediction horizon. The goal is to minimize a cost function that typically includes terms for tracking performance, control effort, and constraints on states and inputs.

**Prediction Horizon:** The prediction horizon defines how far into the future the controller looks when making decisions. It balances performance and computational load.

**Recursion:** NMPC employs a receding horizon strategy. After computing the optimal control actions for the horizon, only the first control input is applied. The process is repeated at the next time step with updated system states.

**Constraints Handling:** One of NMPC's strengths is its ability to handle constraints on both inputs and states explicitly. This is particularly important in practical applications where systems must operate within specific limits.

**Nonlinearity:** Unlike linear MPC, NMPC specifically addresses nonlinearities in the system model, making it suitable for a wider range of applications where linear approximations may not suffice.

### 7.4.2 Potential applications of NMPC

NMPC has been applied to several fields, including:

**Process Control:** Used in chemical and petrochemical processes where nonlinear dynamics are prevalent.

**Robotics:** Helps in motion planning and control for robotic systems, especially in dynamic environments.

**Automotive:** Applied in vehicle dynamics control, including trajectory tracking and path-following problems.

**Aerospace:** Useful in flight control systems where nonlinear effects are significant.

### 7.4.3 Main challenges

The design of NMPC has several challenges, including:

**Computational Complexity:** Solving the optimization problem in real time can be computationally intensive, especially for high-dimensional systems.
**Model Accuracy:** The performance of NMPC heavily relies on the accuracy of the used model of the system. Model mismatch can lead to sub-optimal performance or instability.
**Parameters tuning:** Proper tuning of the cost function and constraints is crucial for achieving desired performance and robustness.

## 7.5 Proposed running controllers

This section addresses the problem of control for planar moving of a Kangaroo hopping robot. Marc Raibert [2] is at the origin of most current controls in the field of hopping robots [25–27]. First, the control approach developed by Raibert for a one-legged planar hopping robot is described. Then, a nonlinear model predictive control approach is proposed to control the leg thrust and is combined with Raibert's controller.

### 7.5.1 Raibert's controller

In the early 1980s, Marc Raibert designed a three-part scheme to control a planar one-legged hopping robot. Its control system deals with thrust, forward speed, and body attitude as three separate control problems.

#### 7.5.1.1 *Thrust control*

The basic principle of this control approach is to apply, by means of the linear actuator, a constant force $F_l$ on the leg during its extension in the stance phase:

$$F_l = cst, \quad \text{if} \quad \dot{l} = \frac{dl}{dt} > 0 \tag{7.4}$$

This control enables us to compensate the energy losses due to friction by injecting the energy during the stance phase and to control the height of the jump.

#### 7.5.1.2 *Forward speed control*

The hopping robot is an underactuated system, therefore it is not possible to control directly the horizontal movement of the robot while controlling the thrust and the body attitude. The position of the foot, when it touches the ground at the end of the flight phase, has a crucial impact on the evolution of the passive part of the robot during the following contact. Thus there is a neutral foot position $x_{f0}$ in relation to the hip position at the touchdown that provides unaccelerated travel for the body of the robot during the stance phase. To accelerate/decelerate the machine, the control system displaces the foot from its neutral position. Acceleration is needed to stabilize the forward speed against errors and external disturbances and to switch from one forward speed to another. The control system uses a linear function of the error in forward speed to find a displacement for the foot:

$$x_f = x_{f0} + k_{\dot{x}}(\dot{x} - \dot{x}_d) \quad \text{with } x_{f0} = \frac{\dot{x} T_s}{2} \tag{7.5}$$

where

$x_f$ is the desired displacement of the foot from the hip at touchdown;
$\dot{x}$ is the forward speed;
$\dot{x}_d$ is the desired forward speed;
$T_s$ is the duration of the stance phase, which depends on the leg stiffness;
$k_{\dot{x}}$ is the velocity feedback gain.

#### 7.5.1.3 *Body attitude control*

The robot's body can be positioned only during the contact phase. The contact between the ground and the foot of the robot in this phase enables the angular actuator of the hip to control the angle of the body without causing high accelerations on the leg. The control system enables us to control the torque thanks to a conventional PD controller and allows the body to follow the desired path:

$$\Gamma = k_p(\phi_d - \phi) - k_v(\dot{\phi}) \tag{7.6}$$

where

$\Gamma$ is the torque exerted by the angular actuator of the hip;
$\phi$ is the body angle between the longitudinal axis of body and the horizontal axis;
$\phi_d$ is the desired body angle relative to the horizontal axis;
$k_p, k_v$ are, respectively, the proportional and derivative gains of the controller.

### 7.5.2 Proposed NMPC running controller

In order to track more effectively the references of speed and jumping height, we propose to improve Raibert's control approach. The improvements concern the thrust exerted by the leg before the take-off. Indeed, the correct control of the leg angle and speed of the robot during the take-off allows us to control the ballistic trajectory of the robot (cf. illustration in Fig. 7.4) in the flight phase. This phase is the longest on a complete jumping cycle, therefore it is crucial to control the speed and the robot's position in this period. The control of the foot placement at touchdown and the behavior of the body during the stance phase are those of Raibert's control scheme (cf. Section 7.5.1).

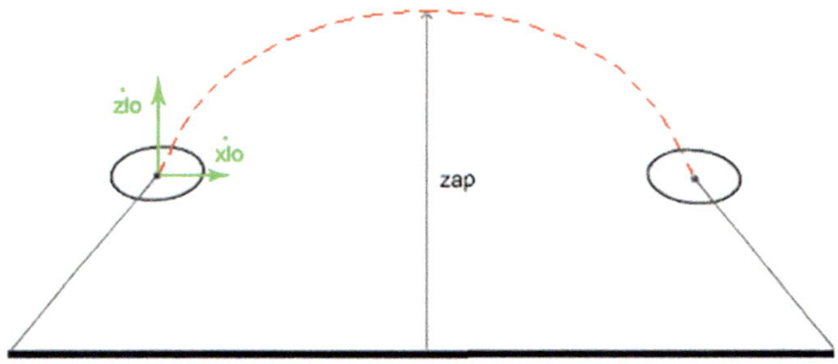


**FIGURE 7.4** Trajectory of the robot during the flight.

### 7.5.2.1 Ballistic trajectory

The ballistic trajectory is determined during lift-off. The force of gravity exerts continuously an action on the vertical acceleration of the robot. Therefore the horizontal component of the velocity of the robot's body $\dot{x}$ does not vary during the flight phase. The horizontal velocity is preserved, this speed has to correspond to the desired speed of movement of the robot at lift-off $\dot{x}_{lo} = \dot{x}_d$. The work of Harbick and Sukhatme [10] has enabled us to determine the vertical velocity of the robot's body needed at the lift-off $\dot{z}_{lo}$ to jump to a desired height $z_{apd}$, given by the following:

$$\dot{z}_{lo} = \sqrt{2g(z_{apd} - z_{lo})} \tag{7.7}$$

where

$\dot{z}_{lo}$ is the vertical velocity at lift-off;
$z_{apd}$ is the desired apex of the jump;
$z_{lo}$ is the body height at lift-off;
$g$ is the acceleration due to gravity.

The vertical and horizontal components of velocity of the robot at lift-off required to track the movement speed and desired jumping height have been determined. To follow these speeds, the corresponding angle and speed of the leg have to be identified. By deriving the coordinates of the center of mass of the robot's body at lift-off, the following set of nonlinear equations is obtained:

$$\begin{cases} \dot{x}_{lo} = -\dot{l}_d \sin(\upsilon_d) - l\dot{\upsilon}\cos(\upsilon_d) \\ \dot{z}_{lo} = \dot{l}_d \cos(\upsilon_d) - l\dot{\upsilon}\sin(\upsilon_d) \end{cases} \tag{7.8}$$

The desired angle $\upsilon_d$ and speed of leg $\dot{l}_d$ to be followed are then identified.

### 7.5.2.2 Basic principle of the proposed control approach

The aim of the model predictive control (also called receding horizon predictive control) is to find a control sequence to minimize a performance criterion $J$

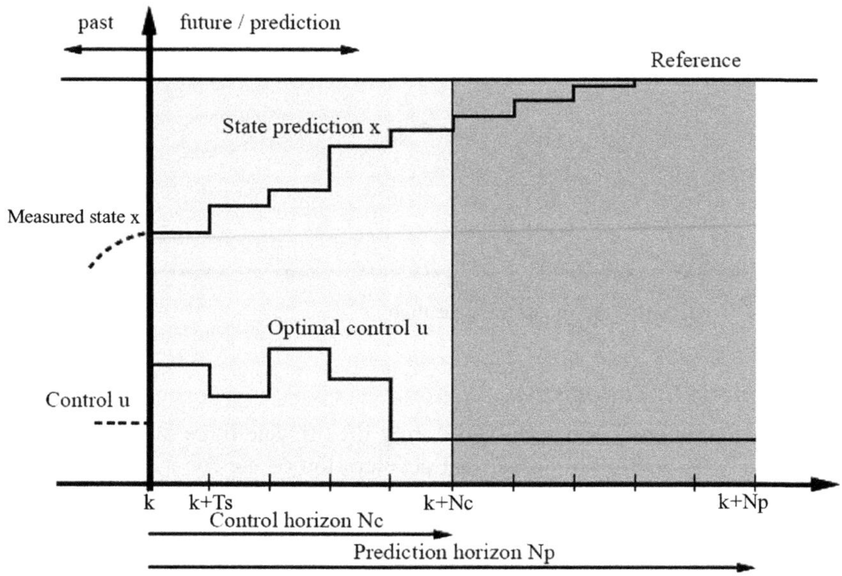


**FIGURE 7.5** Model predictive control (basic principle).

(also called the cost function) subject to constraints:

$$\min_{U} \quad J = \sum_{k=1}^{Np} (x_d - x_k)^T Q (x_d - x_k) + U^T R U \tag{7.9}$$

$$under \ x_k \in C \tag{7.10}$$

where $x_k$ and $x_d$ are, respectively, the state and the desired state of the system. $U$ is the control input and the set $C$ gathers the constraints on the state along the prediction horizon $Np$. $Q$ and $R$ are, respectively, the weighting matrices on the system's state and control input.

This basic principle is illustrated in Fig. 7.5. In the case of constrained nonlinear optimization problems (which is our case), it is impossible to find an analytical solution. Therefore the computer must solve numerically the optimization problem at each sampling time. It must determine a control sequence of size $Nc$, which minimizes the criterion $J$ on a prediction horizon $Np$ taking into account the eventual constraints. The control horizon is generally less than or equal to the prediction horizon [24]. At the end of the control horizon, the control input is kept at zero until the end of the prediction horizon.

From the control sequence computed along the prediction horizon, only the first sample is really applied to the system. Indeed, the application of this control leads to a state that may be different from the predicted one because of modeling errors/uncertainties and the inevitable perturbations that may affect

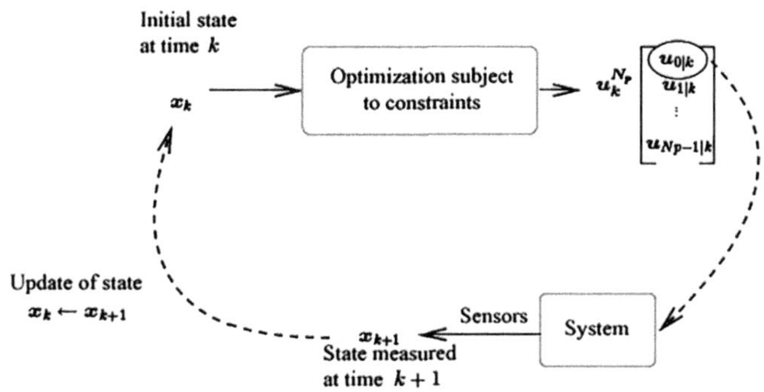


FIGURE 7.6 Model predictive control algorithm.

the system. To overcome this problem, the closing of the loop of the system is necessary (cf. Fig. 7.6). This amount corresponds to applying the technique of receding horizon. Indeed, after applying the first sample of the computed control sequence, the current state of the system is measured, the horizon is shifted by one sampling period and the same optimization problem is restarted considering the current measured state as initial state. This operation is repeated at each sampling instant, which will improve significantly the performance of the control approach and increase its robustness.

Nonlinear model predictive control is a particularly effective scheme for underactuated systems unlike other conventional control approaches that have limitations for controlling this type of system. In addition, predictive control has proved its high performance in terms of robustness towards uncertainties and disturbance rejection. It requires, however, a particularly good model. The drawback of this type of approach is the computation time required to solve the optimization problem. It depends essentially on the control horizon, the complexity of the dynamic model, and to a lesser extent, the prediction horizon. However, nowadays many research efforts have been made to apply this control approach to fast systems.

#### 7.5.2.3 Application to control a Kangaroo robot

A modification of Raibert's controller is proposed by replacing the open-loop control of the thrust of the leg in the stance phase by a nonlinear model predictive controller. Therefore the obtained scheme is composed of two Raibert's controllers that can position the foot at the touchdown and control the body attitude. The pressure exerted by the leg is controlled by the proposed nonlinear model predictive controller. When the robot is in contact with the ground and the leg begins relaxing, the nonlinear model predictive control calculates the force to be applied by the linear actuator until the lift-off instant. The controller

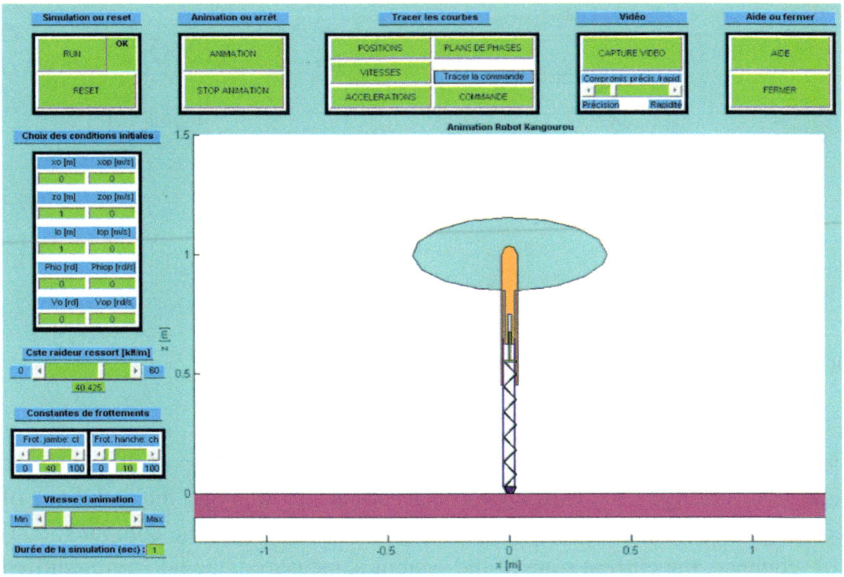


FIGURE 7.7 Graphical user interface of the Kangaroo robot's developed simulator.

minimizes the following performance criterion:

$$\begin{aligned} \underset{u}{min} \quad J &= \sum_{k=1}^{Np} Q_1(l_0(k) - l(k))^2 + Q_2(\dot{x}_{lo}(k) - \dot{x}(k))^2 \\ &\quad + Q_3(\dot{z}_{lo}(k) - \dot{z}(k))^2 \\ under \quad |u_k| &< u_{max} \quad for\ all \quad k \in \{1, ..., N_c\} \end{aligned} \tag{7.11}$$

This optimization criterion is used to follow horizontal speed $\dot{x}_{lo}$ and vertical speed $\dot{z}_{lo}$ corresponding to the desired ballistic trajectory (cf. Section 7.5.2.1). Therefore the height of the jump as well as the speed of the robot are controlled during flight.

## 7.6 Numerical simulations and results

### 7.6.1 Simulation environment

A simulator Kangaroo robot was developed using the Graphical User Interface of Matlab® software. The developed graphical interface of our simulator is shown in Fig. 7.7.

It consists of 11 items:

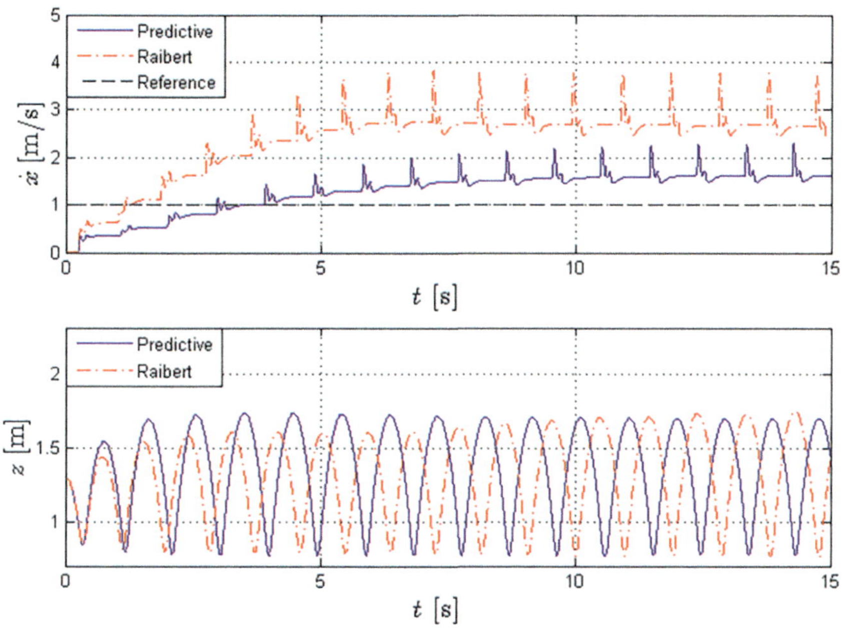


**FIGURE 7.8** Evolution of the horizontal speed and the vertical position of the hip for both proposed controllers.

- ✓ Simulation or reset
- ✓ Animation or stop
- ✓ Plot curves
- ✓ Video capture
- ✓ Help or close
- ✓ Initial conditions
- ✓ Spring stiffness coefficient
- ✓ Friction coefficient
- ✓ Animation speed
- ✓ Duration of simulation
- ✓ Graphical view

This graphical interface enables us to easily manipulate simulations, choose/change initial conditions and parameters of the robot as well as the simulation parameters. It also allows us to display the robot movement during simulation.

### 7.6.2 Simulation comparative study

Both Raibert's control and the nonlinear model predictive control are implemented in our simulator. The same modeling parameters are used to easily compare the performance of these two control schemes. Simulation results obtained are presented in Figs. 7.8–7.11. The different parameters of the dynamic model are summarized in Table 7.1. In Fig. 7.8 and 7.11 dash-dotted lines correspond to Raibert's control, while solid lines correspond to nonlinear model predictive control.

The evolution of horizontal velocity (cf. Fig. 7.8 top) and vertical displacement (cf. Fig. 7.8 lower) shows that the two schemes enable control of the robot

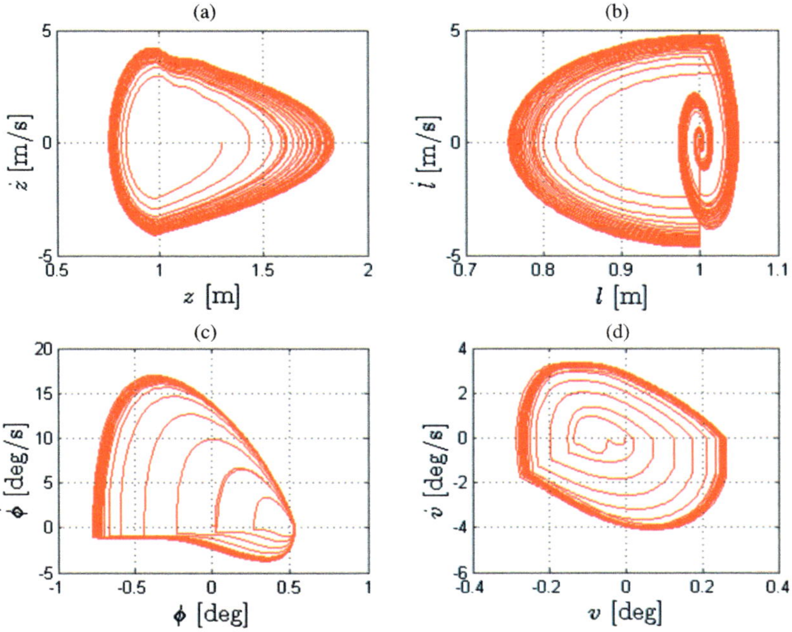


**FIGURE 7.9** Raibert's control: phase portraits.

jumps to converge to cyclic trajectories. The desired reference hopping speed is 1 m/s (cf. dashed lines in Fig. 7.8 top). With the assumption that the robot starts movements from rest the initial speed is fixed at zero, then it converges in about 6 s to 2.7 m/s with Raibert's control and in about 9 s to 1.6 m/s with nonlinear model predictive control. Predictive control converges a little slower than Raibert's control to a steady state, however, it gives a more precise tracking. Indeed, the reference on the hopping speed is much better tracked with the proposed predictive control since the error on the speed is reduced to 60% with respect to 170% with Raibert's control. For the height of the jump (cf. Fig. 7.8 lower), it converges much faster towards the final value with nonlinear model predictive control. The transient regime is about 2 s, while it is about 12 s with Raibert's control. Unlike Raibert's control that uses open-loop control to compute the force exerted on the leg, nonlinear predictive control allows it to follow, in a closed loop, a reference on the height of the jump. In our simulation, a reference of 1.8 m is fairly well followed, the steady state error is about 6%.

Phase portraits $(z, \dot{z})$, $(\phi, \dot{\phi})$, $(l, \dot{l})$, and $(v, \dot{v})$ shown in Fig. 7.9 (for Raibert's control) and Fig. 7.10 (for nonlinear model predictive control) highlight a faster convergence of the controlled system with nonlinear model predictive control. The generated forces and torques exerted, respectively, in the leg and hip are presented in Fig. 7.11. The open-loop control of Raibert (cf. dash-dotted line in Fig. 7.11 top) that controls the thrust of the leg during the stance phase

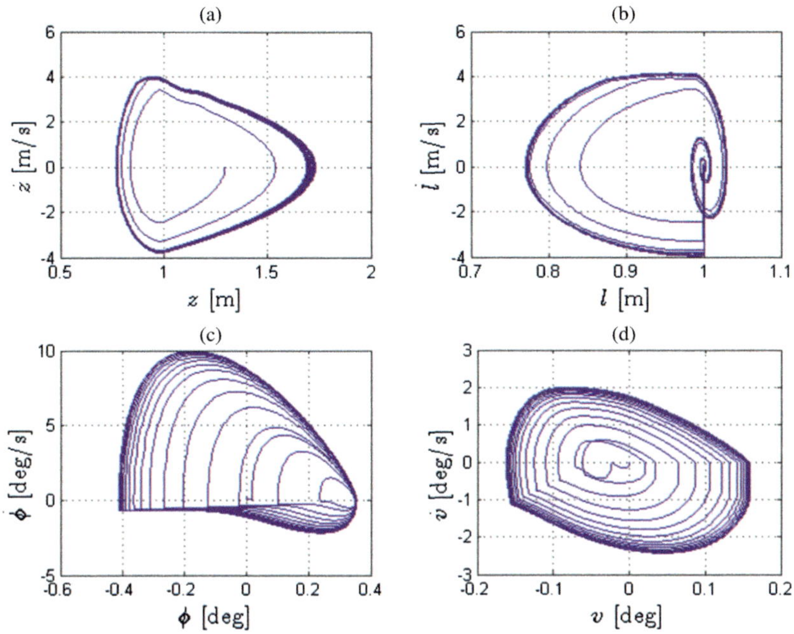


**FIGURE 7.10** Nonlinear model predictive control: phase portraits.

was set to 240 N so that the jumping height converges to 1.8 m. This command does not enable us to control the speed of horizontal movement. The thrust of the leg produces a horizontal speed excessive for the flight phase. During this phase, the forward speed controller positions the robot's leg forward to allow the robot to slow down during the following contact phase (cf. Section 7.5.1.2). Although the mass and inertia of the body are fairly small compared to those of the leg, the torque exerted at the hip (cf. dash-dotted line in Fig. 7.11 lower) causes a rotation of the body as wide as the angular clearance of the leg and will be important. The speed of the robot is much higher than the reference, the body undergoes an important rotation (cf. Fig. 7.9(c)) at touchdown. The control law to reposition the body during the stance phase involves a torque of particular importance at the hip, which can reach 1450 N m.

The nonlinear model predictive control enables us to take into account the constraints of the system to control the force exerted on the leg (cf. solid line Fig. 7.11 top). Here, the control has been deliberately saturated ($U_{max} = 400$ N in (7.11)). Actually, the system will converge more rapidly towards its permanent state as this value is higher. In our case, three or four jumps are necessary to reach the high jump and speed of desired movement. The control of the horizontal speed at lift-off minimizes the angular clearance of the leg and body of the robot during the flight phase (cf. Figs. 7.10(c) and 7.10(d)). Consequently, the torque exerted to control the body's orientation in the sagittal plane during

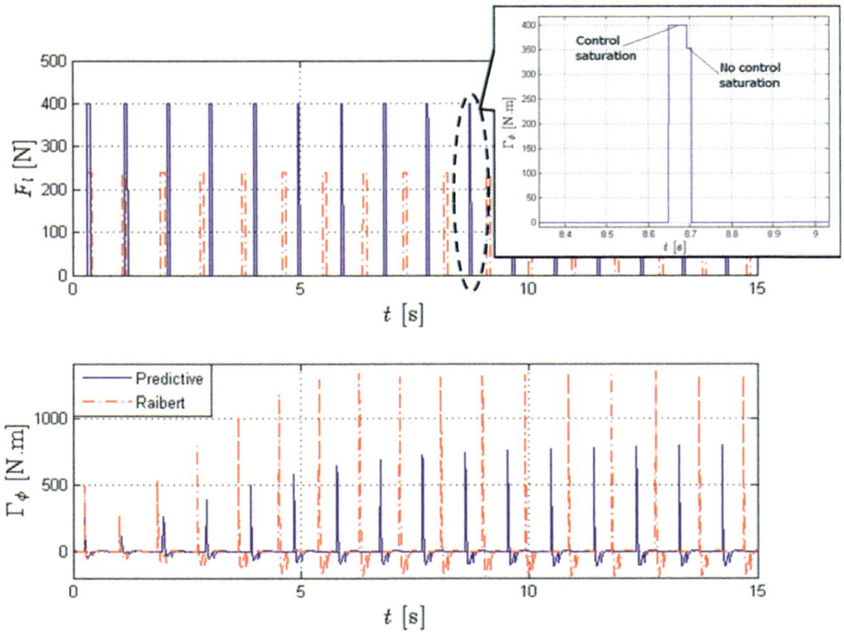


**FIGURE 7.11** Raibert and nonlinear model predictive control: control signals.

the stance phase is less important than with Raibert's controller (cf. Fig. 7.11 lower). To show the saturation of the control at $U_{max}$ and the verification of this constraint (imposed in the optimization criterion (7.11)), a zoom is achieved on the control input computed by the nonlinear predictive controller (cf. Fig. 7.11 and Fig. 7.12.)

The use of the nonlinear predictive scheme to control the thrust before lift-off has significantly improved the overall behavior of the "Kangaroo" robot. Now, it is possible to control the height of jump of the robot, which was impossible with Raibert's controller. The robot follows more precisely references and converges faster to a stable cyclic movement. Furthermore, this control enables us to improve the response of Raibert's forward speed and body attitude control, minimizing the torque exerted at the hip.

## 7.7 Conclusion

The objective of this chapter was to design a control approach for the one-legged planar hopping Kangaroo robot in order to generate stable periodic motions. To reach this objective, the Lagrangian dynamic model of the hopping robot on the different phases of the jumping cycle has been computed and a simulator has been realized. Then, a control scheme has been proposed to generate stable periodic motions in the sagittal plane. The proposed control scheme on the whole jumping cycle is derived from Raibert's control approach. To overcome

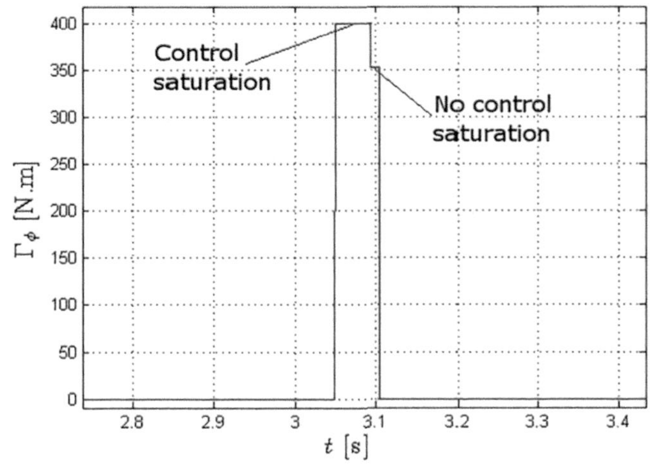


**FIGURE 7.12** Nonlinear model predictive control: zoom on the control signal $F_l$.

the Raibert's control drawbacks related to the movements speed and jumping height a nonlinear model predictive control has been introduced to improve its performances. The obtained simulation results show clearly that the new control approach is much more efficient.

## References


[1] B. Siciliano, L. Sciavicco, L. Villani, G. Oriolo, Robotics: Modelling, Planning and Control, 2009.

[2] M. Raibert, Legged Robots That Balance, Massachusetts Institute of Technology, 1986.

[3] M. Ishikawa, A. Neki, J. Imura, S. Hara, Energy preserving control of a hopping robot based on hybrid port-controlled Hamiltonian modeling, in: SICE 2003 Annual Conference (IEEE Cat. No.03TH8734), vol. 3, 2003, pp. 2611–2616.

[4] K. Matsuoka, A mechanical model of repetitive hopping movements, Biomechanisms 5 (1979) 251–258.

[5] H. Taghirad, Analysis, Design, and Control of Hopping Robot, Master thesis, 1993.

[6] N. Cherouvim, E. Papadopoulos, Energy saving passive-dynamic gait for a one-legged hopping robot, Robotica 24 (2006) 491–498.

[7] M. Ahmadi, M. Buehler, Stable control of a simulated one-legged running robot with hip and leg compliance, IEEE Transactions on Robotics and Automation 13 (1997) 96–104.

[8] K.G. Naik, M. Mehrandezh, J.M. Barden, Control of a one-legged hopping robot using a hybrid neuro-PD controller, IEEE Canadian Conference On Electrical And Computer Engineering (2006) 1530–1533.

[9] H. Rad, P. Gregorio, M. Buehler, Design, modeling and control of a hopping robot, IEEE/RSJ Conference Intelligent Systems And Robots 3 (1993) 1778–1785.

[10] K. Harbick, G. Sukhatme, Height control for a one-legged hopping robot using a one-dimensional model, Technical Report IRIS-02-413, Institute for Robotics, 2001.

[11] K. Harbick, G. Sukhatme, Controlling hopping height of a pneumatic monopod, IEEE International Conference On Robotics And Automation 4 (2002) 3998–4003.

[12] W. Schwind, D. Koditschek, Control of forward velocity for a simplified planar hopping robot, IEEE International Conference On Robotics And Automation 1 (1995) 691–696.

[13] N. Cherouvim, E. Papadopoulos, Single actuator control analysis of a planar 3DOF hopping robot, Robotics: Science And Systems (2005) 145–152.
[14] A. Sato, Simulation of one-legged hopping robot with phase plane stability, in: IASTED International Conference: Modelling and Simulation, 2007, pp. 112–117.
[15] A. Sato, M. Buehler, A planar hopping robot with one actuator: design, simulation, and experimental results, IEEE/RSJ International Conference On Intelligent Robots And Systems 4 (2004) 3540–3545.
[16] G. Zeglin, B. Brown, Control of a bow leg hopping robot, IEEE International Conference On Robotics And Automation 1 (1998) 793–798.
[17] Ö. Arslan, U. Saranli, Ö. Morgül, Reactive footstep planning for a planar spring mass Hopper, in: IEEE/RSJ International Conference Intelligent Robots and Systems, 2009, pp. 160–166.
[18] S. Hyon, T. Mita, Development of a biologically inspired hopping robot-"Kenken", IEEE International Conference on Robotics and Automation 4 (2002) 3984–3991.
[19] M. Kovac, D. Floreano, A miniature 7g jumping robot, in: IEEE International Conference on Robotics and Automation, 2008, pp. 373–378.
[20] S. Blank, T. Wahl, T. Luksch, K. Berns, Biologically inspired compliant control of a monopod designed for highly dynamic applications, in: IEEE/RSJ International Conference on Intelligent Robots and Systems, 2009, pp. 148–153.
[21] K. Hosoda, H. Takayama, T. Takuma, Bouncing monopod with bio-mimetic muscular-skeleton system, in: IEEE/RSJ International Conference on Intelligent Robots and Systems, 2008, pp. 3083–3088.
[22] M. Spong, Underactuated mechanical systems, in: Control Problems in Robotics and Automation, 1998.
[23] M. Spong, M. Vidyasagar, Robot Dynamics and Control, John Wiley & Sons, 1989.
[24] E. Camacho, C. Bordons, Model Predictive Control, 2nd edition, Springer, 2007.
[25] K. Saar, F. Giardina, F. Iida, Model-free design optimization of a hopping robot and its comparison with a human designer, IEEE Robotics and Automation Letters 3 (2018) 1245–1251.
[26] T. Verstraten, R. Furnémont, P. Beckerle, B. Vanderborght, D. Lefeber, A hopping robot driven by a series elastic dual-motor actuator, IEEE Robotics and Automation Letters 4 (2019) 2310–2316.
[27] X. Yu, F. Iida, Minimalistic models of an energy-efficient vertical-hopping robot, IEEE Transactions on Industrial Electronics 61 (2014) 1053–1062.
[28] S. Upadhyay, A. Aguiar, Constrained hopping traversability analysis on non-uniform polygonal chains, IEEE Access 8 (2020) 36691–36701.

Chapter 8

# Dual model-free control

**Contents**



## 8.1 Introduction

Model-free control strategies have been proposed in [1,2] resulting in a breakthrough in nonlinear control. This technique is based on the results on fast estimation and identification of nonlinear signals [3,4]. The control scheme is based on a local linear approximation of the controlled system dynamics that is valid for a small time window. This approximation is updated in an online fashion thanks to a fast estimator. The control law proposed consists in a PID controller augmented with compensating terms provided by the online estimation of the system dynamics. The overall controller is also called an *i*-PID (intelligent PID) controller. Comparison of such a controller with a classical PID controller can be found in [5]. The main advantage of this control strategy is that it does not require either prior knowledge of the system dynamics, or complex parameter tuning. It is therefore easy to build a controller for an unknown system.

Model-free control has been successfully applied to many academic control problems as well as various industrial cases. Linear and nonlinear systems are studied in [1,6,7] as well as the ball and beam mechanical system for both stabi-

Control of Underactuated Mechanical Systems. https://doi.org/10.1016/B978-0-44-324020-1.00017-8 

lization and reference trajectories tracking in [7]. Application to switched nonlinear systems, which is a generalization of hybrid systems, is studied in [2]. In [8], a motor's throttles are regulated using a model-free controller. Shape Memory Alloys (SMA) are able to modify their shape when heated. They are used to design compact actuators but their dynamics and therefore their associated control problem remains complex. In [9,10], model-free control is successfully used to control a SMA-based actuator. Water level control in open channels is often subject to unpredictable disturbances of significant magnitudes. In [11], water level control for hydroelectric power plants is achieved using model-free control techniques, providing a single control law for a wide range of flow. In [12], model-free control methodology is applied to a power converter, where stable regulation is achieved for large variations of current intensity. Modern financial engineering involves tracking control of "risk-free" management. For this application, in [13,14], model-free control is compared to other existing techniques, showing the superiority of the model-free control scheme. Further recent applications can be found in [16–22].

Although this control method is smart, and has been applied to resolve many control problems, it has some drawbacks. In addition to its dependency on the quality of sensors and sampling frequency, on which relies the fast local estimation, model-free control is not currently adapted to control of non-minimum-phase systems. Those systems are characterized with unstable internal dynamics. In this chapter we focus on stable limit cycles generation for non-minimum-phase systems, more precisely for underactuated mechanical systems. A mechanical system with fewer actuators than degrees of freedom is said to be underactuated and unfortunately, a vast majority of these systems are non-minimum-phase ones [23–26]. Therefore model-free control techniques cannot be applied as it has initially been proposed. Some efforts have been made for two particular cases of non-minimum-phase systems: the ball and beam [7] (where the dynamics of the beam has not been taken into account) and the Planar Vertical Take Off and Landing (PVTOL) aircraft [15].

In order to achieve stable limit cycles on all coordinates of a non-minimum-phase underactuated system, we first design a family of parametrized periodic trajectories for a subset of coordinates. Those trajectories are then tracked using control inputs thanks to a classical model-free technique. Since the system is non-minimum phase, the internal dynamics of the system is unstable. Therefore stable limit cycles on those coordinates are generated through the control of trajectory parameters. To achieve this control, we use another model-free controller using untracked coordinates as output and trajectory parameters as input. The proposed methodology is illustrated through two examples of underactuated mechanical systems with numerical simulations, namely the cart-pole pendulum and the pendubot. Numerical simulations show the effectiveness and the robustness of the proposed control approach towards external disturbances and changes in system dynamics.

## 8.2 Background on model-free control

Model-free control is based on a local linear approximation of the input–output behavior of the system to be controlled. This approximation is valid over a short period of time. For simplification purposes, the model-free control approach is presented in the case of a single-input–single-output (SISO) nonlinear system with one control input and one output. However, the method remains applicable to multiple-input–multiple-output (MIMO) systems [1].

### 8.2.1 Nonlinear system to be controlled

Consider a nonlinear system with unknown dynamics. The input–output behavior of this system can be expressed in the following general form:

$$E(y, \dot{y}, \ldots, y^{(a)}, u, \dot{u}, \ldots, u^{(b)}) = 0 \tag{8.1}$$

where $u$ is the control input and $y$ is the system output. Provided this finite-dimensional ordinary differential equation (ODE) is smooth enough, it can be approximated for a short period of time by the following simplified model:

$$y^{(\nu)} = F + \alpha u \tag{8.2}$$

The order of derivation $\nu$ and the constant parameter $\alpha \in \mathbb{R}$ can be chosen arbitrarily. In the literature on model-free control, $\nu$ is generally chosen equal to 1 or 2. The constant $\alpha$ is a design parameter. $F \in \mathbb{R}$ captures all nonlinearities in the input–output behavior and can be compensated for in the control law. Since Eq. (8.2) is only valid over a small time window, it must be corrected at each sampling step. For this reason, the value of $F$ is updated from measurements of the $\alpha u$ and $y^{(\nu)}$ terms as follows:

$$[F(k)]_e = [y^{(\nu)}(k)]_e - \alpha u(k-1) \tag{8.3}$$

where $[F(k)]_e$ is the estimate of the value of $F$ at sampling time $k$ that will be used to calculate the control input at the same sampling time $u(k)$. $[y^{(\nu)}(k)]_e$ is the estimated value of the $\nu$th derivative of the output $y$ at sampling time $k$ and $u(k-1)$ is the control input used at the previous sampling time $k-1$ (the value of $F$ can be initialized to 0 at time $k=0$).

### 8.2.2 Control law

Given the estimate of $F$ expressed by Eq. (8.3), the control law simply compensates for unknown nonlinearities and adds a control term corresponding to a closed-loop tracking $y^*(t)$ reference trajectory in the form of a conventional PID controller. The resulting controller is called an *intelligent*-PID (*i*-PID):

$$u = \frac{1}{\alpha}\left(-F + y^{*(\nu)} + K_p e + K_i \int e(t)\,\mathrm{d}t + K_d \dot{e}\right) \tag{8.4}$$

where $y^{*(\nu)}$ is the $\nu$th derivative of the reference trajectory $y^*$, $K_p$, $K_i$, $K_d$ are the PID gains and $e = y^* - y$ is the output tracking error and $\dot{e}$ its first derivative.

The closed-loop system is obtained by substituting the value of the control input (8.4) into the local model (8.2):

$$y^{(\nu)} = y^{*(\nu)} + K_p e + K_i \int e(t)\,\mathrm{d}t + K_d \dot{e} \tag{8.5}$$

The PID gain tuning can therefore be performed by pole placement, under the assumption that nonlinearities are compensated for by the control law. There is therefore no need for complex system identification, as is the case when a conventional PID is used to control a nonlinear system [6,7].

If $\nu = 1$ the PID controller reduces to a PI controller:

$$u = \frac{1}{\alpha}\left(-F + \dot{y^*} + K_p e + K_i \int e(t)\,\mathrm{d}t\right) \tag{8.6}$$

and the closed-loop dynamics becomes:

$$\dot{y} = \dot{y^*} + K_p e + K_i \int e(t)\,\mathrm{d}t \tag{8.7}$$

## 8.3 Proposed dual model-free control solution

In our case, we are particularly interested[1] in the generation of stable limit cycles on underactuated mechanical systems that are generally non-minimum phase. Once again, for the sake of simplicity, the control is applied to a one-input control system with two degrees of freedom, which are the minimum dimensions required for a system to be underactuated. We recall the dynamics of an underactuated mechancial system in the general case:

$$m_{11}\ddot{q}_a + m_{12}\ddot{q}_{na} + h_1 + g_1 = u \tag{8.8}$$

$$m_{21}\ddot{q}_a + m_{22}\ddot{q}_{na} + h_2 + g_2 = 0 \tag{8.9}$$

where $q = [q_a, q_{na}]^T$ is the generalized coordinate vector ($q_a$ is the actuated coordinate and $q_{na}$ is the non-actuated coordinate) and $u$ is the control input. The state of the system is assumed to be the vector $[q_a\ q_{na}\ \dot{q}_a\ \dot{q}_{na}]^T$.

### 8.3.1 Basic principle

The aim of the control is to generate stable limit cycles on both the actuated and non-actuated coordinates. To this end, we define a family of reference trajectories $q^*_{na}(t, p)$, parameterized by a parameter $p \in \mathcal{P}$ and periodic with period $t_c$ on the non-actuated coordinate (cf. Section 8.4.1.2).

[1] The proposed approach here has been developed during the Ph.D thesis of Sébastien Andary to whom we are grateful for permitting us to reproduce some of his results.

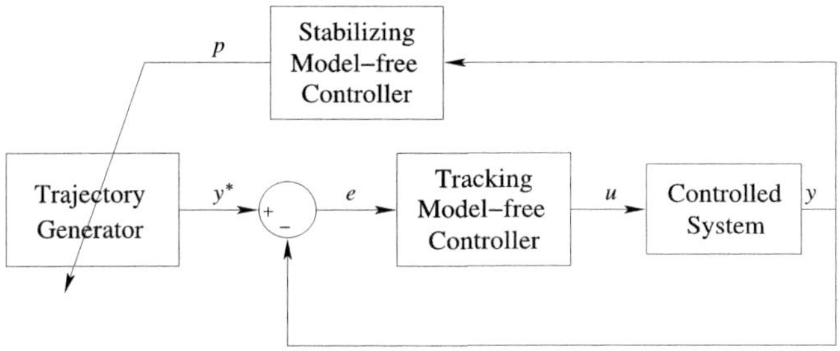


**FIGURE 8.1** Block diagram of the dual model-free controller. The system has $y$ as output and $u$ as input. Reference trajectories for the output $y^*$ are generated for a given value of parameter $p$, then tracked by a first model-free controller (i.e., tracking controller) using the output error $e = y^* - y$. The parameter $p$ is updated by a second model-free controller (i.e., a stabilization controller).

A first model-free controller is used to track reference trajectories on the non-actuated coordinate using the $u$ control input. In order to stabilize the internal dynamics of the resulting closed-loop system, (i.e., flywheel dynamics) and generate stable limit cycles on both coordinates, a second model-free controller is used. The latter therefore has the $p$ parameter of the reference trajectories as its control variable, and uses the actuated coordinate as the control variable. At the end of each cycle of the periodic reference trajectory, the second controller calculates the $p$ parameter used to stabilize the actuated coordinate. The $p$ parameter selected then completely defines the reference trajectory to be used by the first controller during the next cycle. A block diagram representing a general view of the proposed controller is shown in Fig. 8.1.

### 8.3.2 Control design

The design of the reference trajectory tracking controller is based on model-free control. The dynamics of the non-actuated coordinate is approximated by the local model following the concept of model-free control:

$$\ddot{q}_{na} = F_1 + \alpha_1 u \tag{8.10}$$

where the constant $\alpha_1$ is a design parameter, and $F_1$ is a variable term intended to capture nonlinearities in the dynamics of the non-actuated coordinate and is updated according to Eq. (8.3) at each sampling period. The controller tracking the reference trajectories is obtained using the numerically calculated value of $F_1$, resulting in an intelligent PID ($i$-PID):

$$u = \frac{1}{\alpha_1}\left(-F_1 + \ddot{q}^*_{na}(p,t) + K_{p1}e + K_{i1}\int e(t)\,\mathrm{d}t + K_{d1}\dot{e}\right) \tag{8.11}$$

With appropriately chosen gains $K_{p1}$, $K_{i1}$, and $K_{d1}$, the unactuated coordinate follows the desired periodic trajectory $q^*_{na}(p,t)$.

The $p$ parameter used in the control law (8.11) is constant over half a period, i.e., $\forall t \in \left[k\frac{t_c}{2}\ (k+1)\frac{t_c}{2}\right[$ $(k \in \mathbb{N})$ and only changes value at the end of each half-period at instants $(k+1)\frac{t_c}{2}$ thanks to the second controller (8.14). In addition, the nonlinear dynamics of the actuated coordinate is approximated by the following local discrete model:

$$\Delta v_a(k) = F_2(k) + \alpha_2 p(k) \tag{8.12}$$

where $\Delta v_a(k) = \dot{q}_a\left(k\frac{t_c}{2}\right) - \dot{q}_a\left((k-1)\frac{t_c}{2}\right)$ is the angular velocity variation of the actuated coordinate $v_a(k) = \dot{q}_a\left(k\frac{t_c}{2}\right)$ measured at the limits of a half-period, and the constant $\alpha_2$ is a design parameter. The value of $F_2$ is updated at instants $t_k = k\frac{t_c}{2}$ for $k \in \mathbb{N}$, at the end of each half-period, according to the principle of model-free control:

$$[F_2(k)]_e = [\Delta v_a(k)]_e - \alpha_2 p(k-1) \tag{8.13}$$

The notation $[.]_e$ stands for the estimated value of [.] as explained in Section 8.2. This discretization is justified as the aim of the second controller is to generate a stable limit cycle on the actuated coordinate. To achieve this, it is sufficient for the system's actuated coordinate to return to a fixed state $(q_a^d, \dot{q}_a^d)$ at the end of each cycle. Only the effect of parameter $p$ on the state of the system at the end of each half-period is therefore taken into account.

The second model-free controller calculates every half-period the reference trajectory parameter $p$ according to the following formula:

$$\begin{aligned} p(k) = \frac{1}{\alpha_2}\Bigg( & - F_2(k) + \Delta v_a^d(k) + K_{p2}\left[q_a^d\left(k\frac{t_c}{2}\right) - q_a\left(k\frac{t_c}{2}\right)\right] \\ & + K_{i2}\int q_a^d\left(k\frac{t_c}{2}\right) - q_a\left(k\frac{t_c}{2}\right) \mathrm{d}t + K_{d2}\left[\dot{q}_a^d\left(k\left(\frac{t_c}{2}\right) - \dot{q}_a\left(k\frac{t_c}{2}\right)\right)\right]\Bigg) \end{aligned} \tag{8.14}$$

where $K_{p2}$, $K_{i2}$, $K_{d2}$ are the PID gains. Note that since the desired state for the actuated coordinate $(q_a^d, \dot{q}_a^d)$ is fixed, the term $(\Delta v_a)^d$ is zero.

## 8.4 Application to underactuated mechanical systems

The application of this control scheme is illustrated through two examples, including (i) the classical cart-pole inverted pendulum, and (ii) the pendubot. In these examples, the aim of the control is to generate stable limit cycles on all system coordinates.

### 8.4.1 Application 1: The cart-pole inverted pendulum

The cart-pole inverted pendulum (cf. Fig. 8.2) is an underactuated mechanical system often used in the control theory literature. Its dynamics is nonlinear,

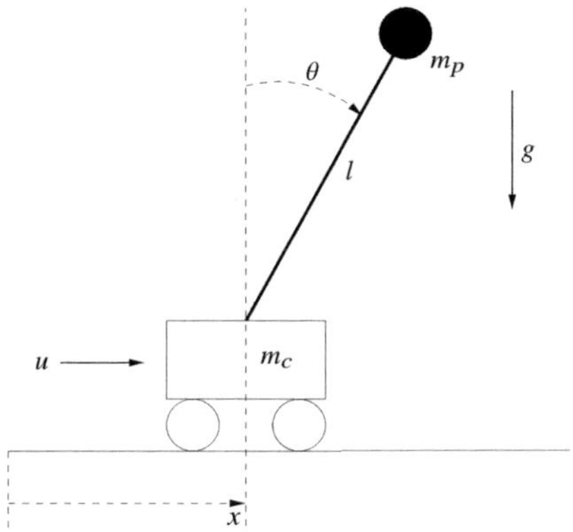


**FIGURE 8.2** Schematic view of the cart-pole inverted pendulum. The variable $\theta$ represents the angle between the pendulum and the vertical and $x$ is the position of the carriage on the rail. The control input $u$ is the force applied to the carriage.

but simple to model and therefore easy to simulate. The system consists of a pendulum attached to a carriage through a passive joint. The carriage can move linearly on a rail in the same plane as the pendulum link, and is driven by a motor.

#### 8.4.1.1 Dynamic model

The control input $u$ is the horizontal force applied to the carriage and the degrees of freedom are the angular position $\theta$ of the pendulum relative to the vertical and the linear position $x$ of the carriage on the rail. Fig. 8.2 shows the principle of the inverted pendulum on a carriage.

The nonlinear model governing the dynamics and movements of this system is as follows [27]:

$$(m_p + m_c)\ddot{x} + m_p l \cos\theta \ddot{\theta} - m_c \dot{\theta}^2 \sin\theta = u \tag{8.15}$$

$$m_p l \cos\theta \ddot{x} + m_p \ddot{\theta} - m_p l g \sin\theta = 0 \tag{8.16}$$

where $m_p$ and $m_c$ are the masses of the pendulum and carriage, respectively, $l$ is the length of the pendulum and $g$ is the acceleration due to gravity. Table 8.1 shows the dynamic parameters of the system used in the simulations.

#### 8.4.1.2 Reference trajectories generation

In the application examples of this control method, the choice was made to construct periodic trajectories (oscillatory in form), cut into half-periods and then generated over a complete cycle, based on the property of symmetry between two half-periods with respect to the time axis. The parametrization of these trajectories should allow the controller to update the parameter $p$ (which corresponds to the time at which the trajectory $q_{na}^*$ crosses zero) during tracking

**TABLE 8.1 Summary of dynamic parameters of the inverted pendulum.**

| Parameter | Description | Value |
|---|---|---|
| $m_c$ | mass of the carriage (kg) | 0.3 |
| $m_p$ | mass of the pendulum (kg) | 0.01 |
| $l$ | length of the pendulum (m) | 1 |
| $g$ | acceleration due to gravity (m s$^{-2}$) | 9.81 |

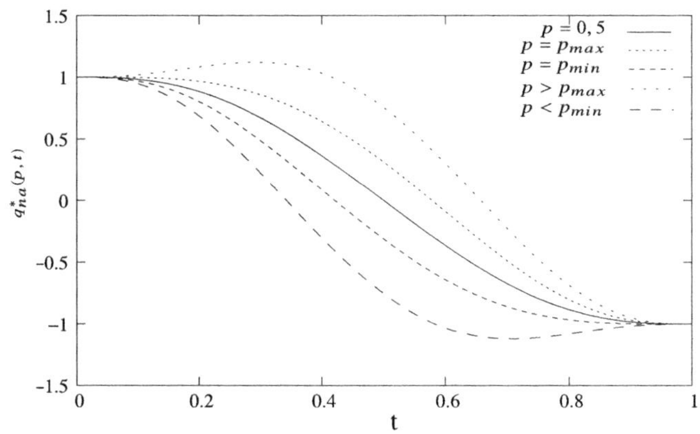


**FIGURE 8.3** Evolution of half-period normalized reference trajectories for the non-actuated coordinate $q^*_{na}(p,t)$ for different values of $p$. The period $t_c$ is equal to 2 s and the amplitude $2A$ is equal to 2.

while the overall trajectory remains smooth. These considerations bring us to the initial and final conditions of each half-period. Given the period $t_c$ of the oscillations and their amplitude $2A$, the boundary conditions are as follows:

$$\forall p \in \mathcal{P}, \begin{cases} q^*_{na}(p,0) = q^*_{na}(p,t_c) = A \\ q^*_{na}(p,\frac{t_c}{2}) = -A \\ \dot{q}^*_{na}(p,0) = \dot{q}^*_{na}(p,\frac{t_c}{2}) = \dot{q}^*_{na}(p,t_c) = 0 \end{cases} \tag{8.17}$$

for a certain range $\mathcal{P} \subset \mathbb{R}$ of parameter $p$ values. A polynomial function of degree six satisfying these constraints is then used, parameterized by $p$ such that:

$$\forall p \in \mathcal{P},\ q^*_{na}(p,t=p) = 0 \tag{8.18}$$

The calculation of this polynomial is detailed in the next sections. Fig. 8.3 shows the normalized reference trajectories ($t_c = 2$, $A = 1$) used for the examples for different values of the parameter $p \in [p_{min},\ p_{max}]$ during a half-period. The $\mathcal{P}$ domain is restricted to the interval $[p_{min},\ p_{max}]$ to ensure that the trajectory remains in the interval $[-A,\ A]$ for all $t \in [0,\ \frac{t_c}{2}]$ and thus retains a sinusoidal shape (see next sections). Note that $p_{min} = 1 - p_{max}$ due to the symmetry property of half-period sections. The trajectory generation parameters for this example are chosen as follows: the trajectory amplitude is $A = 22.5°$ and the period is $t_c = 2$ s.

#### 8.4.1.3 Control law

For these simulations, the control parameters used are: $\alpha_1 = 10$, $\alpha_2 = -1$. The gains of the first controller are set to $K_{p1} = 10$, $K_{i1} = 0$, and $K_{d1} = 2$. The gains of the second controller are $K_{p2} = 0, 2$, $K_{i2} = 0$, and $K_{d2} = 1$. The second controller requires a desired state $(x^d, \dot{x}^d)$ around which to regulate the actuated coordinate. Since the dynamics of the system is unknown when the model-free controller is designed, we assume that we do not know the position of the cart at the end of each half-period to fix $x^d$. This is because, even in steady state, the oscillations of the carriage are such that at the end of each half-period, the position of the cart alternates between two extreme positions. It is therefore not possible to select its desired position $x^d$ for the second controller in Eq. (8.14) a priori. Instead of using the $x(k\frac{t_c}{2})$ measurement, we use the average of the last two consecutive measured values of the carriage position, allowing us to choose the average value of the desired position $x^d_{avg}$ as 0 and the desired carriage speed $\dot{x}^d$ as 0 in order to obtain oscillations of the carriage around the origin.

#### 8.4.1.4 Simulation results

In numerical simulations, an external disturbance is considered in the form of a torque applied pointwise to the pendulum at time $t = 20$ s, with an intensity of 0.1 N m. The results obtained for this simulation are shown in Figs. 8.4 and 8.5. Convergence to a stable limit cycle is clearly visible on the pendulum and carriage phase portraits. The disturbance is slightly visible on the pendulum trajectory in the form of a deviation from the reference trajectories. The response of the first controller can be observed in the evolution of the control input. The torque generated almost instantaneously compensates for the effect of the disturbance on the pendulum's trajectory. The effect of this disturbance is greater on the cart trajectory. The second controller adapts the reference trajectories to limit the effect of the disturbance and returns the trajectories to the limit cycle within a few periods. The second controller's response is not instantaneous, as its control law is only updated at the end of each half-period.

### 8.4.2 Application 2: The pendubot

The proposed control approach is also applied to the pendubot [27] that was presented in Section 6.2.4. Also often used in the literature on control of underactuated mechanical systems. Shown in Fig. 6.3, it is a planar manipulator arm with two degrees of freedom, where the first link is actuated, while the second is passive.

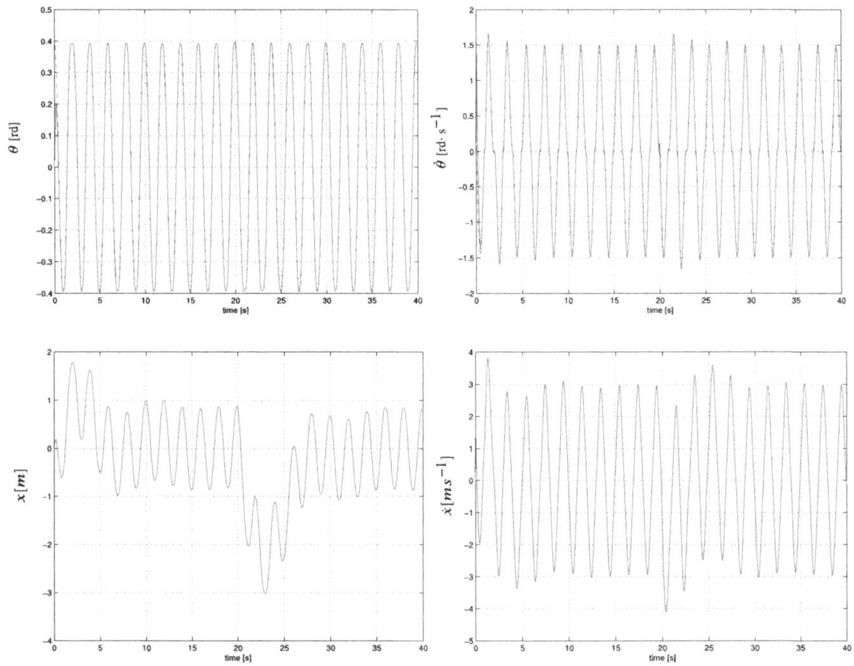


**FIGURE 8.4** Simulation results (system states) of the inverted pendulum on cart. A point disturbance is introduced in the form of a torque applied to the pendulum at time $t = 20$ s.

### 8.4.2.1 Dynamic model

The system control input $u$ represents the torque applied to the first body, and the degrees of freedom are the angular position $q_1$ of the first body relative to the vertical and the angular position $q_2$ of the second body relative to the first. In this scenario, we are interested in controlling the absolute angular positions of the two bodies relative to the vertical, $q_1$ and $(q_1 + q_2)$.

Let us recall the nonlinear dynamic model of the system used for the control design as well as numerical simulations [27]:

$$m_{11}\ddot{q}_1 + m_{12}\ddot{q}_2 + h_1 + g_1 = u \tag{8.19}$$

$$m_{21}\ddot{q}_1 + m_{22}\ddot{q}_2 + h_2 + g_2 = 0 \tag{8.20}$$

Table 8.2 summarizes the dynamic parameters of the pendubot used for numerical simulations.

### 8.4.2.2 Reference trajectories generation

The control aims to generate stable limit cycles on the two absolute angular positions $q_1$ and $(q_1 + q_2)$. The reference trajectories are generated following the method proposed in Section 8.4.1.2 and tracked on the $(q_1 + q_2)$ coordinate. The amplitude of these trajectories is $A = 3°$ and their period is $t_c = 2$ s.

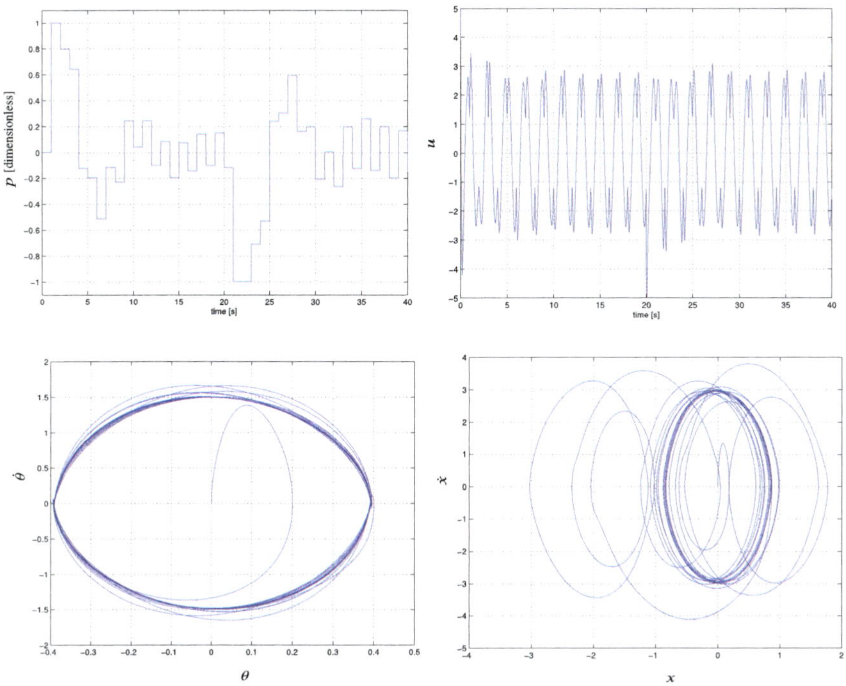


**FIGURE 8.5** Simulation results (optimization parameter, control input, and phase portraits) of the inverted pendulum on a cart. A point disturbance is introduced in the form of a torque applied to the pendulum at time $t = 20$ s.

**TABLE 8.2 Summary of dynamic parameters of the pendubot.**

| Parameter | Description | Value |
|---|---|---|
| $m_1$ | mass of the first rod (kg) | 1 |
| $m_2$ | mass of the second rod (kg) | 1 |
| $I_1$ | inertia of the first rod (kg m$^2$) | 0.0833 |
| $I_2$ | inertia of the second rod (kg m$^2$) | 0.0833 |
| $l_{c1}$ | distance to the center of mass of the first rod (m) | 0.5 |
| $l_{c2}$ | distance to the center of mass of the second rod (m) | 0.5 |
| $l_1$ | length of the first rod (m) | 1 |

### *8.4.2.3 Control law*

In this simulation, the control parameters used for the first controller are $\alpha_1 = -10$, $K_{p1}$ =50, $K_{i1} = 0$, and $K_{d1}$ =10. The parameters of the second controller are $\alpha_2 = 0, 15$, $K_{p2} = 2$, $K_{i2} = 0$, and $K_{d2} = 1$. For the same reasons as in Section 8.4.1, the average of the last two consecutive values of $q_1(k\frac{t_c}{2})$ is used in

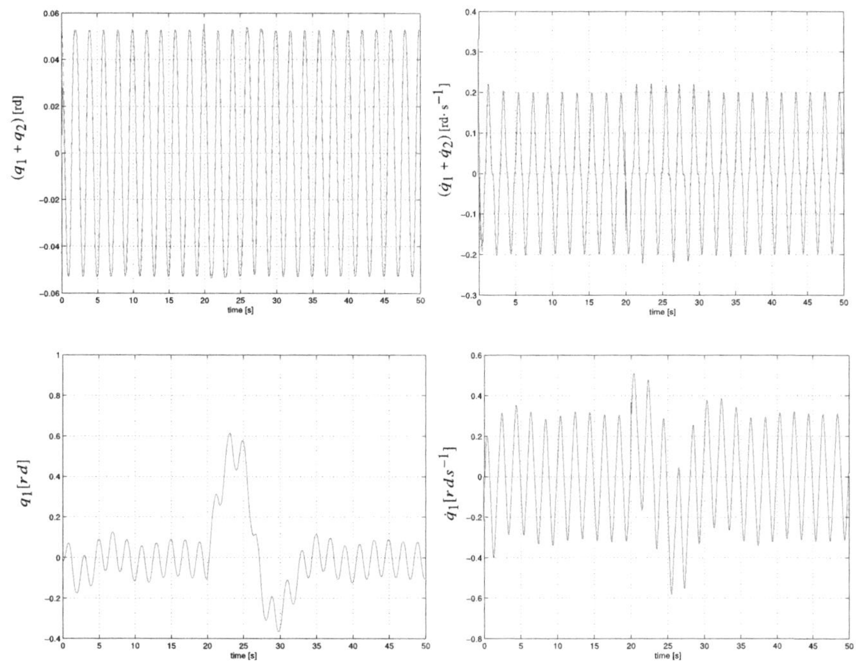

**FIGURE 8.6** Simulation results (system states) of the pendubot. A point disturbance is introduced in the form of a torque applied on the passive joint at time $t = 20$ s.

the second controller. The desired mean value $q_{1}{}_{avg}^{d}$ is 0 and the desired angular velocity $\dot{q}_1^d$ is 0.

#### 8.4.2.4 Simulation results

In this simulation, an external disturbance is introduced through a torque applied punctually to the second joint at time $t = 20$ s with an intensity of 1 N m. The results are shown in Figs. 8.6 and 8.7. Limit cycles are obtained on each coordinate (actuated and non-actuated) of the system, and can be observed in the respective phase portraits. The first controller's response to the external disturbance limits its effect on trajectory tracking to a small deviation in position and velocity of the $q_1 + q_2$ output. The second controller, on the other hand, adjusts the reference trajectory parameter and brings the trajectories of the actuated coordinate back toward the limit cycle within a few periods.

### 8.4.3 Application 3: The inertia wheel inverted pendulum

Thanks to the dynamic coupling between the two pendulum coordinates, the non-actuated coordinate (the pendulum angle) can be directly controlled via the control input (the torque applied to the flywheel). Reference trajectories can therefore be followed on the non-actuated coordinate. In the system dynamic

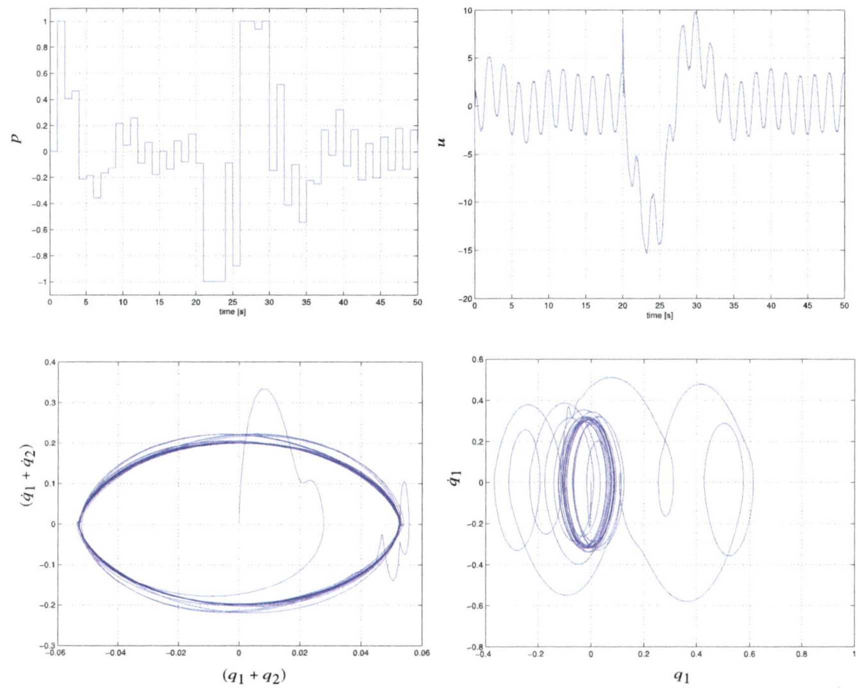

FIGURE 8.7 Simulation results (optimization parameter, control input, and phase portraits) of the pendubot. A point disturbance is introduced in the form of a torque applied on the passive joint at time $t = 20$ s.

Eq. (6.67), the disturbance torque $C_1$ is assumed to be zero and the torque $C_2$ (generated by the actuator) is the control input $\tau$ generated by the proposed control scheme. The system equation can be simplified as follows:

$$\ddot{\theta}_1 = \frac{1}{I}\left(-\tau + \overline{ml}g\sin\theta_1\right) \tag{8.21a}$$

$$\ddot{\theta}_2 = \frac{I + i_2}{Ii_2}\tau - \frac{\overline{ml}g}{I}\sin\theta_1 \tag{8.21b}$$

According to the basic principle of model-free control, Eq. (8.21a) for the dynamics of the non-actuated coordinate $\theta_1$ is approximated by the following differential equation:

$$\ddot{\theta}_1 = F_1 + \alpha_1\tau \tag{8.22}$$

where $\alpha_1$ is a control parameter chosen by the designer. The term $F_1$ is estimated at each sampling time using the control at the previous sampling time and the estimated acceleration:

$$[F_1(k)]_e = [\ddot{\theta}_1(k)]_e - \alpha_1\tau(k-1) \tag{8.23}$$

with $[\ddot{\theta}_1(k)]_e$ the estimated acceleration of the pendulum angle.

#### 8.4.3.1 Control law

The model-free control law for tracking reference trajectories on the non-actuated coordinate $\theta_1$ is as follows:

$$\tau = \frac{1}{\alpha_1}\left(-F_1 + \ddot{\theta}_1^{ref}(t, p) + K_{p1}e + K_{i1}\int e(t)\,\mathrm{d}t + K_{d1}\dot{e}\right) \tag{8.24}$$

where $e = \theta_1^{ref} - \theta_1$ is the tracking error, $\dot{e}$ is its derivative. $K_{p1}$, $K_{i1}$, and $K_{d1}$ represent the gains of the associated PID. This control is completely defined, given the reference trajectory $\theta_1^{ref}(t, p)$ and its derivatives.

The second model-free controller in the proposed solution is responsible for adjusting the $p$ parameter of the reference trajectories every half-period. The command calculated by this controller represents the $p$ parameter of the reference trajectory.

The closed-loop dynamics of the actuated coordinate (the flywheel) is approximated by the following discrete system:

$$\Delta_{t_c} v_{\theta_2} = F_2 + \alpha_2 p \tag{8.25}$$

where $\Delta_{t_c} v_{\theta_2} = \dot{\theta}_2(k\frac{t_c}{2}) - \dot{\theta}_2((k-1)\frac{t_c}{2})$ is the variation in flywheel speed $v_{\theta_2} = \dot{\theta}_2$ measured between two half-periods. $\alpha_2$ is a control setting parameter and $F_2$ is estimated at the end of each half-period as follows:

$$[F_2(k)]_e = [\Delta_{t_c} v_{\theta_2}(k)]_e - \alpha_2 p(k-1) \tag{8.26}$$

where $[\Delta_{t_c} v_{\theta_2}(k)]_e$ is the estimated variation in flywheel speed measured between half-periods $(k-1)$ and $k$ and $p(k-1)$ the value of the control input (parameter $p$) during the previous half-period.

The purpose of this controller is to obtain a zero flywheel speed at the end of each half-cycle, in order to converge the actuated coordinate $\theta_2$ towards a stable limit cycle. The output of the desired actuated coordinate is therefore set to $\dot{\theta}_2^d = 0$ (the angular position of the flywheel is of no interest and is therefore not taken into account).

The second model-free controller calculates the $p$ parameter of the reference trajectories for the next half-period as follows:

$$p = \frac{1}{\alpha_2}\left(-F_2 + K_{d2}(\dot{\theta}_2^d - \dot{\theta}_2)\right) \tag{8.27}$$

with $K_{d2}$ the PID gain. It is worth noting that this control law is simplified compared to the general case presented in this chapter. Since the angular position of the flywheel is not taken into account, the proportional and integral gains are zero in this case.

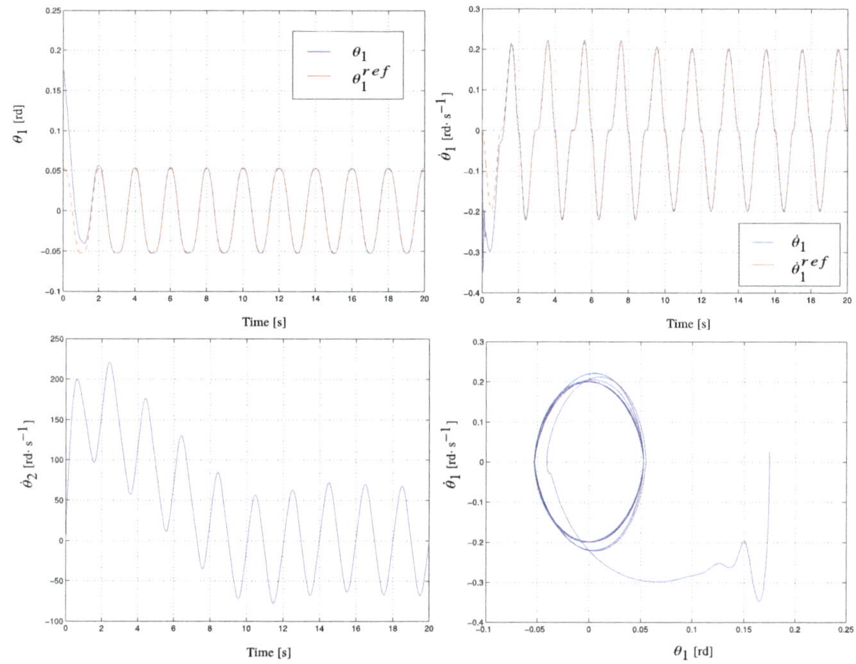


**FIGURE 8.8** Simulation results of scenario 1: Nominal case (positions, speeds, and phase portrait) of the inertia-wheel inverted pendulum.

### *8.4.3.2 Simulation results*

The reference trajectory tracking controller parameters used are: $\alpha_1 = -100$, $K_{p1} = 200$, $K_{i1} = 0$, and $K_{d1} = 100$. The reference trajectory used is that described above with amplitude $A = 3°$ and period $\tau = 2$ s. The reaction wheel stabilization controller is set to $\alpha_2 = 50$, $K_{p2} = K_{i2} = 0$, and $K_{d2} = 1$. The sampling frequency is set to 150 Hz.

Two simulation scenarios are considered. In the first scenario, the system undergoes no external disturbance except for non-zero initial conditions. In the second scenario, an external disturbance is considered in the form of a disturbance torque $C_1 \neq 0$ applied to the pendulum, in order to test the controller's ability to reject this disturbance.

## Scenario 1: Nominal case

In this simulation, no external disturbances are introduced apart from non-zero initial conditions. Initial conditions are chosen, as in previous simulations, according to the position of the pendulum at rest (resting on the mechanical stop). The results of this scenario are shown in Figs. 8.8 and 8.9. The pendulum trajectories are shown in Figs. 8.8 (top left) and (top right). The pendulum's real trajectory (solid line) quickly joins the reference trajectory (dashed line). The

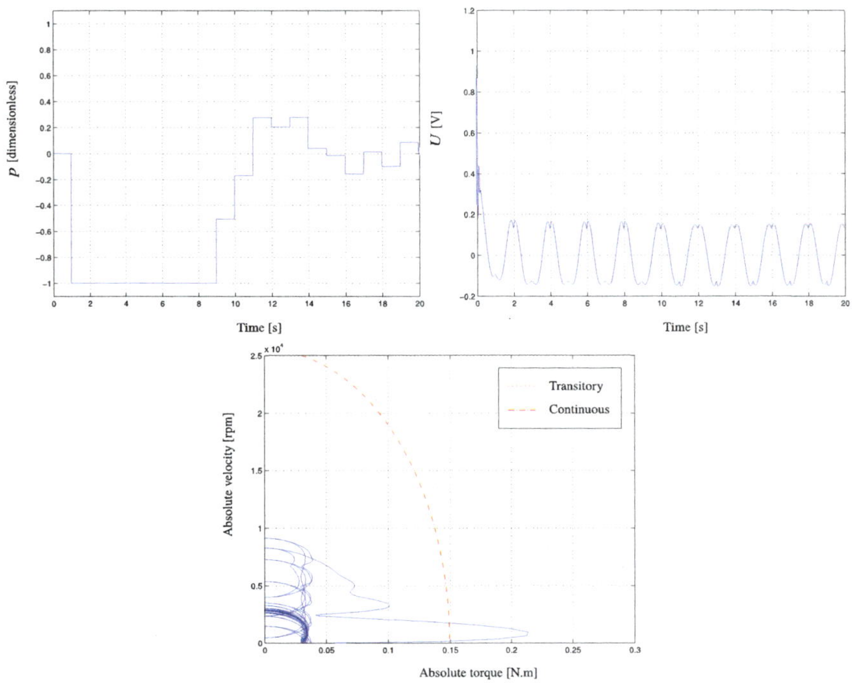


**FIGURE 8.9** Simulation results of scenario 1: Nominal case (optimization parameter, control input, and power admissibility) of the inertia-wheel inverted pendulum.

evolution of the flywheel velocity is shown in Fig. 8.8 (lower left). Non-zero initial conditions cause the trajectory to deviate from the desired state ($\dot{\theta}_2 = 0$). However, the second controller brings this trajectory back to the desired limit cycle around zero. The phase plane of the pendulum, plotted in Fig. 8.8 (lower right), shows the convergence of the pendulum's trajectory towards a stable limit cycle.

We can observe the evolution of the two command inputs $u$ and $p$ from the two controllers, respectively in Figs. 8.9 (top right) and (top left). The second controller deforms the reference trajectories to bring the steering wheel speed trajectory back to a stable limit cycle. Finally, Fig. 8.9 (lower) shows the absolute flywheel speed as a function of the absolute motor torque as well as the admissible actuator power region, and shows that the trajectory remains within these acceptable limits.

## Scenario 2: External punctual disturbance rejection

In this scenario, an external disturbance is introduced at time $t = 15$ s in the form of a torque applied to the pendulum (represented by $C_1$ in the system dynamics model) with an intensity of 0.5 N m. The aim is to observe the system's behavior towards point disturbances. Initial conditions are the same as in the

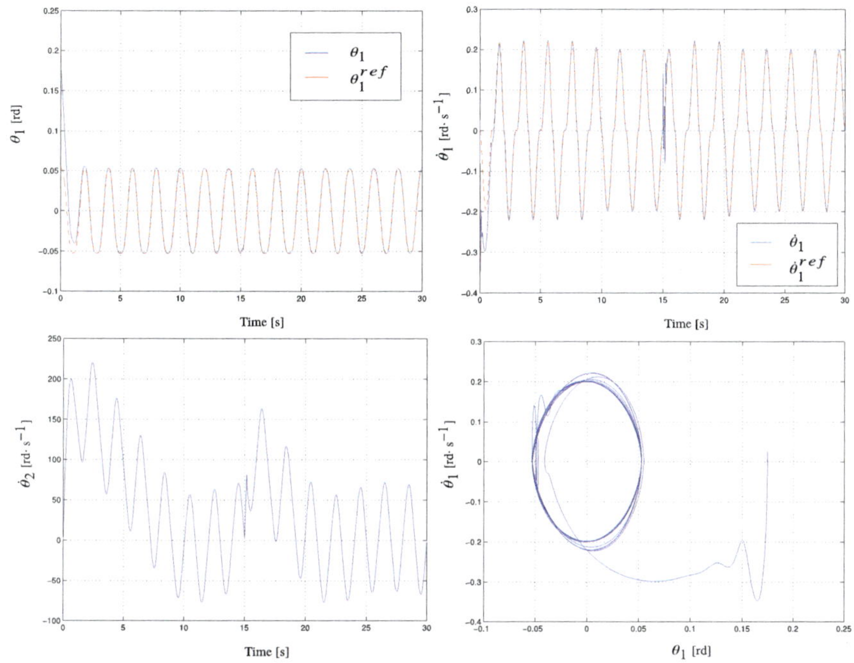


**FIGURE 8.10** Simulation results of scenario 2: Punctual external disturbance rejection (positions, speeds, and phase portrait) of the inertia-wheel inverted pendulum.

previous scenario. The obtained results are shown in Figs. 8.10 and 8.11. The pendulum trajectories are plotted in Figs. 8.10 (top left) for position, and 8.10 (top right) for velocity. The effect of the perturbation on the pendulum's position is visible in the form of a slight deviation between the actual trajectory and the reference trajectory. On the other hand, the perturbation is more visible on the pendulum's velocity, where its effect can be seen in the form of a peak.

Fig. 8.10 (lower left) shows the evolution of the flywheel velocity over time. The disturbance causes a consequent shift in the oscillating trajectory, induced by the rapid response of the first controller. However, the flywheel velocity trajectory reaches the limit cycle within a few periods, thanks to the second controller in the proposed approach. The difference in response speed observed between the two controllers is due to the difference in their sampling frequencies. While the first controller adjusts the signal at each sampling instant (i.e., 150 Hz), the second controller adjusts the reference trajectory optimization parameter for each half-period (i.e., 1 Hz). The phase plane of the pendulum is shown in Fig. 8.10 (lower right). Note that the trajectory follows a limit cycle despite the disturbance introduced. The control inputs for the two controllers are shown in Figs. 8.11 (top left) and 8.11 (top right). The reaction of the controllers after the disturbance can be clearly seen. Fig. 8.11 (lower) shows the motor power and its admissible region.

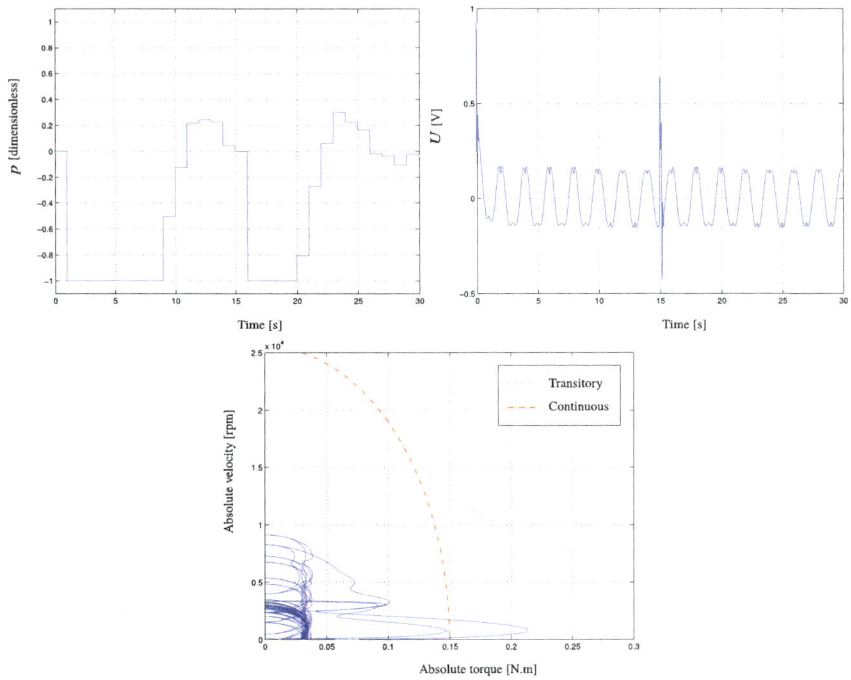


**FIGURE 8.11** Simulation results of scenario 2: Punctual external disturbance rejection (optimization parameter, control input, and power admissibility) of the inertia-wheel inverted pendulum.

### *8.4.3.3 Real-time experimental results*

The proposed control approach has also been implemented on the experimental platform of the inertia wheel inverted pendulum. The controller parameters are different from those used in the numerical simulations. The parameters of the first controller are set to $\alpha_1 = -150$, $K_{p1} = 70$, $K_{i1} = 0$, and $K_{d1} = 12$. The parameters of the second controller are set to $\alpha_2 = 70$, $K_{p2} = K_{i2} = 0$, and $K_{d2} = 1$.

Two experimental scenarios have been conducted. In the first, no external disturbances are deliberately introduced. The only disturbances are non-zero initial conditions and measurement noise. In the second scenario, a point disturbance is introduced by pushing the pendulum sideways during oscillation. The results obtained for these two scenarios are presented below.

#### Scenario 1: Nominal case

In this first scenario, no voluntary external disturbance is applied. Measurement noise and non-zero initial conditions are the only sources of external disturbances. As the control system does not use any model of the system, it can be

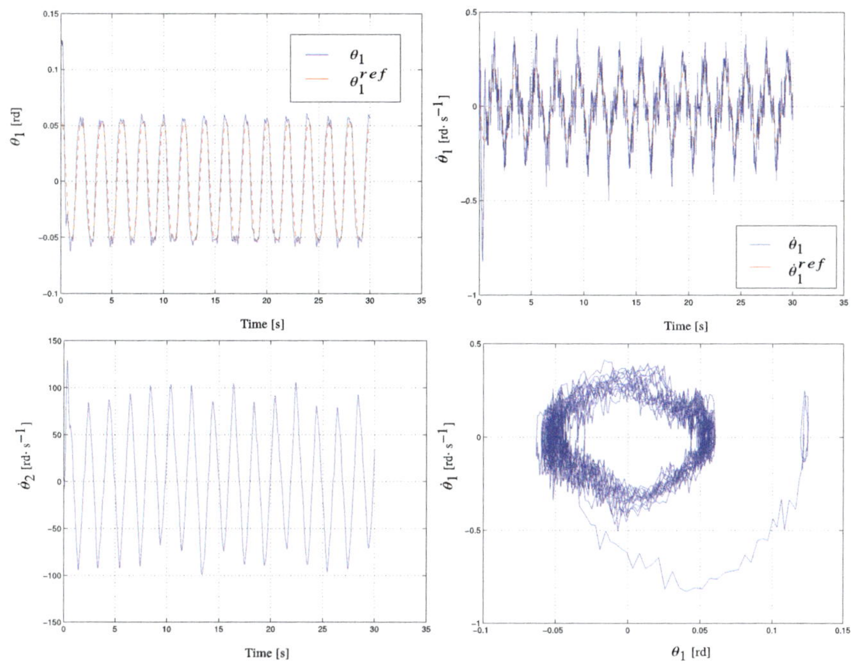


**FIGURE 8.12** Real-time experimental results of scenario 1: Nominal case (positions, speeds, and phase portrait) of the inertia-wheel inverted pendulum.

noted that the disturbances usually induced by modeling and the uncertainty of mechanical parameters are completely compensated for.

Figs. 8.12 and 8.13 illustrate the results of this first experimental scenario. The position and velocity of the pendulum are shown in Figs. 8.12 (top left) and 8.12 (top right), respectively, along with their reference trajectories. Measurement noise is clearly visible on each of the figures, particularly on the pendulum velocity trajectory, where the noise is amplified by the calculation of the numerical derivative. The first controller manages to stabilize pendulum movements around the reference trajectories. Fig. 8.12 (lower left) shows the flywheel velocity trajectory oscillating around the limit cycle thanks to the second controller. The phase portrait of the pendulum is shown in Fig. 8.12 (lower right). The limit cycle is clearly visible, despite the measurement noise. Figs. 8.13 (top right) and 8.13 (top left) show, respectively, the control inputs of the two controllers. Note that the second control input $p$ oscillates around a value different from 0. This offset is used to compensate for persistent disturbances due to poor calibration of the pendulum position sensor. Finally, Fig. 8.13 (lower) demonstrates that the actuator power remains within the admissible range given by the motor manufacturer.

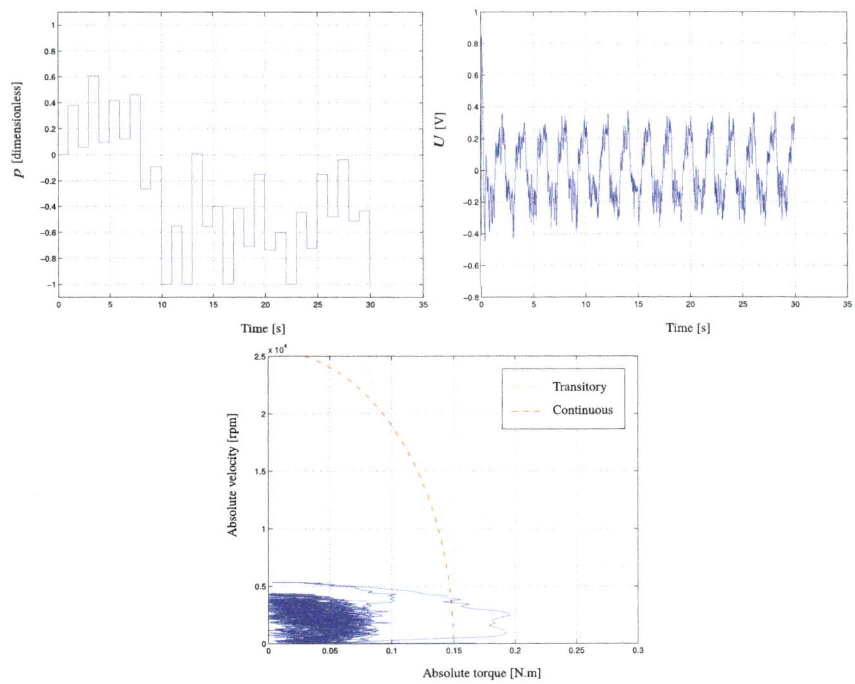


**FIGURE 8.13** Real-time experimental results of scenario 1: Nominal case (optimization parameter, control input, and power admissibility) of the inertia-wheel inverted pendulum.

## Scenario 2: Punctual external disturbance rejection

In the second scenario, two external point disturbances are applied to the pendulum body at instants $t = 12$ s and $t = 23$ s. These disturbances are in addition to those induced by measurement noise and non-zero initial conditions. The parameters of the two controllers, for this scenario, are kept the same as those used in the previous scenario.

The obtained results are shown in Figs. 8.14 and 8.15. Figs. 8.14 (top left) and 8.14 (top right) show the actual trajectories (in terms of position and velocity) as solid lines and the desired trajectories (also in terms of position and velocity) as dashed lines. The effect of the two point disturbances is clearly visible on these trajectories. The flywheel speed is depicted in Fig. 8.14 (lower left), after each disturbance, it reaches the limit cycle. The phase portrait of the pendulum, shown in Fig. 8.14 (lower right), clearly shows convergence towards a stable limit cycle. The $u$ and $p$ control inputs of the two model-free controllers are shown in Figs. 8.15 (top right) and 8.15 (top left), respectively. Finally, Fig. 8.15 (lower) demonstrates that the actuator power remains within the admissible region. These results demonstrate the effectiveness of model-free controllers in stabilizing the system around stable limit cycles, as well as their ability to reject external disturbances.

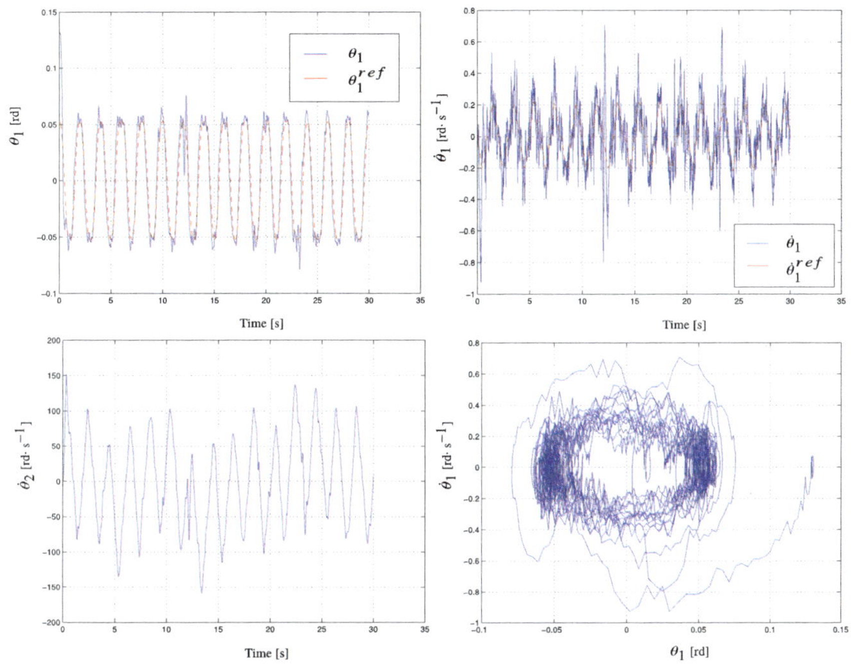


**FIGURE 8.14** Real-time experimental results of scenario 2: Punctual external disturbance rejection (positions, speeds, and phase portrait) of the inertia-wheel inverted pendulum.

## 8.5 Conclusion

This chapter presents an adaptation of model-free control to the special case of generating stable limit cycles on a non-minimum-phase underactuated mechanical system. Model-free control is presented for the general case of a fully actuated system. A dual approach is then proposed to extend this control method to the case of underactuated (non-minimum-phase) mechanical systems. This method inherits the advantages of model-free control, i.e., no mathematical modeling of the controlled system is required, and no complex identification is needed to set the controller parameters.

To illustrate the effectiveness of the proposed control method, three applications of underactuated mechanical systems are proposed: the cart-pole inverted pendulum, the pendubot, and the inertia wheel inverted pendulum. Numerical simulations, as well as real-time experiments show the effectiveness of the proposed control approach as well as its robustness towards external disturbances for the three examples.

The simulations use the inertia wheel inverted pendulum model previously described. Several simulation scenarios are presented. External disturbances are introduced in the form of non-zero initial conditions, and in the form of a disturbance torque applied pointwise to the pendulum. The results attest to the

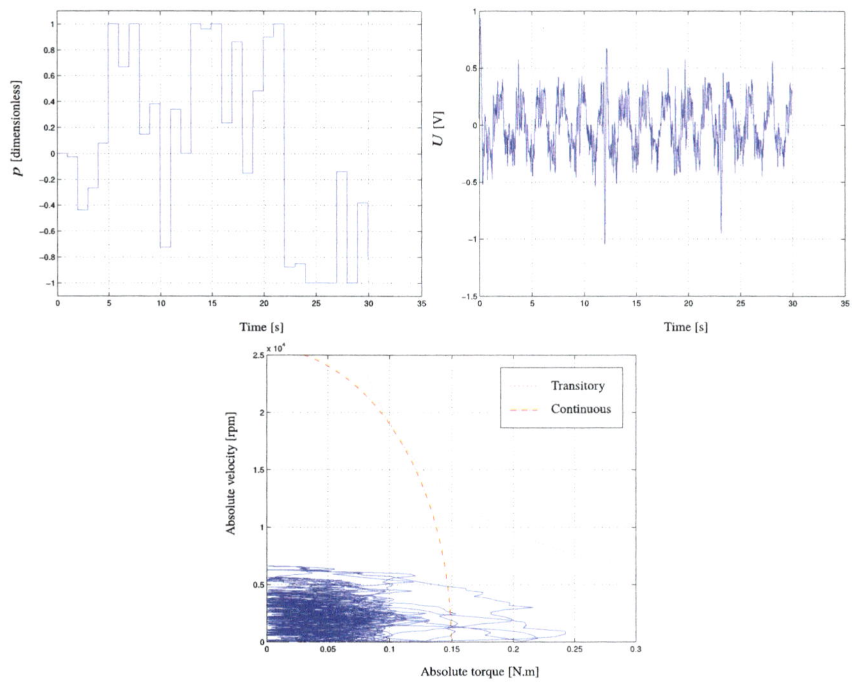


**FIGURE 8.15** Real-time experimental results of scenario 2: Punctual external disturbance rejection (optimization parameter, control input, and power admissibility) of the inertia-wheel inverted pendulum.

effectiveness and feasibility of the proposed control method, and demonstrate its ability to reject external disturbances.

## References


[1] M. Fliess, C. Join, M. Mboup, H. Ramirez, Vers une commande multivariable sans modèle, in: Conférence Internationale Francophone D'automatique (CIFA 2006), 2006.

[2] R. Bourdais, M. Fliess, C. Join, W. Perruquetti, Towards a model-free output tracking of switched nonlinear systems, in: NOLCOS 2007–7th IFAC Symposium on Nonlinear Control Systems, 2007.

[3] M. Fliess, C. Join, H. Ramirez, Non-linear estimation is easy, International Journal of Modelling, Identification and Control 4 (2008) 12–27.

[4] M. Mboup, C. Join, M. Fliess, Numerical differentiation with annihilators in noisy environment, Numerical Algorithms 50 (2009) 439–467.

[5] B. Novel, M. Fliess, C. Join, H. Mounier, B. Steux, A mathematical explanation via "intelligent" PID controllers of the strange ubiquity of PIDs, in: 18th Mediterranean Conference on Control and Automation, MED'10, 2010.

[6] M. Fliess, C. Join, Intelligent PID controllers, in: 16th Mediterrean Conference on Control and Automation, 2008.

[7] M. Fliess, C. Join, Model-free control and intelligent PID controllers: towards a possible trivialization of nonlinear control?, in: 15th IFAC Symposium on System Identification (SYSID 2009), 2009.

[8] C. Join, J. Masse, M. Fliess, Etude préliminaire d'une commande sans modèle pour papillon de moteur, Journal Européen Des Systèmes Automatisés 42 (2008) 337–354.
[9] P. Gédouin, C. Join, E. Delaleau, J. Bourgeot, S. Chirani, S. Calloch, Model-free control of shape memory alloys antagonistic actuators, in: 17th IFAC World Congress, 2008.
[10] P. Gédouin, C. Join, E. Delaleau, J. Bourgeot, S. Chirani, S. Calloch, A new control strategy for shape memory alloys actuators, in: 8th European Symposium on Martensitic Transformations, 2009.
[11] C. Join, G. Robert, M. Fliess, Model-free based water level control for hydroelectric power plants, in: IFAC Conference on Control Methodologies and Technologies for Energy Efficiency, CMTEE, 2010.
[12] L. Michel, C. Join, M. Fliess, P. Sicard, A. Chériti, Model-free control of dc/dc converters, in: 12th IEEE Workshop on Control and Modeling for Power Electronics (COMPEL), 2010.
[13] M. Fliess, C. Join, Delta hedging in financial engineering: towards a model-free approach, in: 18th Mediterranean Conference on Control and Automation, MED'10, 2010.
[14] M. Fliess, C. Join, A model-free approach to delta hedging.
[15] S. Riachy, M. Fliess, C. Join, J. Barbot, Vers une simplification de la commande non linéaire: l'exemple d'un avion à décollage vertical, in: Sixième Conférence Internationale Francophone D'Automatique, CIFA 2010, 2010.
[16] M. Fliess, C. Join, An alternative to proportional-integral and proportional-integral-derivative regulators: intelligent proportional-derivative regulators, International Journal of Robust and Nonlinear Control 32 (2022) 9512–9524.
[17] B. Selma, E. Bounadja, B. Belmadani, B. Selma, M. Fliess, A novel intelligent control approach for wind energy conversion systems with synchronous reluctance generators, International Journal of Circuit Theory and Applications 52 (2024) 3967–3987.
[18] O. Bara, M. Fliess, C. Join, J. Day, S. Djouadi, Toward a model-free feedback control synthesis for treating acute inflammation, Journal of Theoretical Biology 448 (2018) 26–37.
[19] O. Barth, J. Condomines, J. Moschetta, A. Cabarbaye, C. Join, M. Fliess, Full model-free control architecture for hybrid UAVs, in: 2019 American Control Conference (ACC), 2019, pp. 71–78.
[20] M. Fliess, C. Join, Machine learning and control engineering: the model-free case, in: Proceedings of the Future Technologies Conference (FTC) 2020, vol. 1, 2021, pp. 258–278.
[21] L. Michel, I. Neunaber, R. Mishra, C. Braud, F. Plestan, J. Barbot, X. Boucher, C. Join, M. Fliess, Model-free control of the dynamic lift of a wind turbine blade section: experimental results, Journal of Physics. Conference Series 2265 (2022) 032068.
[22] Q. Guilloteau, B. Robu, C. Join, M. Fliess, E. Rutten, O. Richard, Model-free control for resource harvesting in computing grids, in: 2022 IEEE Conference on Control Technology and Applications (CCTA), 2022, pp. 384–390.
[23] R. Marino, P. Tomei, A class of globally output feedback stabilizable nonlinear nonminimum phase systems, IEEE Transactions on Automatic Control 50 (2005) 2097–2101.
[24] X. Hu, Y. Zhao, B. Xu, C. Hu, Robust adaptive fuzzy tracking control for uncertain MIMO nonlinear nonminimum phase system, IEEE Transactions On Systems, Man, And Cybernetics: Systems 50 (2020) 2017–2028.
[25] S. Nazrulla, H. Khalil, Robust stabilization of non-minimum phase nonlinear systems using extended high-gain observers, IEEE Transactions on Automatic Control 56 (2011) 802–813.
[26] M. Benosman, F. Liao, K. Lum, J. Wang, Nonlinear control allocation for non-minimum phase systems, IEEE Transactions on Control Systems Technology 17 (2009) 394–404.
[27] M.W. Spong, Underactuated mechanical systems, in: B. Siciliano, K.P. Valavanis (Eds.), Control Problems in Robotics and Automation, in: Lecture Notes in Control and Information Sciences, vol. 230, Springer, Berlin, Heidelberg, 1998.

# Index